MARINE BIOLOGY

THE FUTURE OF AQUACULTURE FEEDS

MARINE BIOLOGY

Additional books in this series can be found on Nova's website under the Series tab.

Additional e-books in this series can be found on Nova's website under the e-book tab.

MARINE BIOLOGY

THE FUTURE OF AQUACULTURE FEEDS

ELLA LINTON
EDITOR

nova publishers
New York

For permission to use material from this book please contact us:
Telephone 631-231-7269; Fax 631-231-8175
Web Site: http://www.novapublishers.com

Library of Congress Cataloging-in-Publication Data

ISBN: 978-1-63463-069-6

Published by Nova Science Publishers, Inc. † New York

CONTENTS

PREFACE

Chapter 1 - Fish meal and fish oil are important components in the feeds for many farm-raised species, from pigs and poultry to farmed fish. As ingredients in aquaculture feed, fish meal and fish oil supply essential amino acids and fatty acids required for normal growth for cultured species, including carp, salmon, tilapia, trout, catfish, shrimp and others. Fish meal and oil also help maintain the important human health benefits of seafood. However, the relatively high cost of fish meal and fish oil – and growing pressure on the wild fisheries that supply the fish meal and fish oil – are adding up to make alternative feeds one of the top issues facing the global aquaculture industry, fueling research on suitable alternative feed ingredients.

This report provides a comprehensive review of aquaculture feeds and the challenges and opportunities associated with development of various alternatives while suggesting areas for future research.

Chapter 2 – Report of the National Oceanic and Atmospheric Administration on Feeds for Aquaculture: Frequently Asked Questions.

In: The Future of Aquaculture Feeds
Editor: Ella Linton
ISBN: 978-1-63463-069-6

Chapter 1

THE FUTURE OF AQUAFEEDS*

Michael B. Rust, Fredric T. Barrows, Ronald W. Hardy, Andrew Lazur, Kate Naughten and Jeffrey Silverstein

FOREWORD

The National Oceanic and Atmospheric Administration (NOAA) and the Agricultural Research Service and National Institute of Food and Agriculture of the U.S. Department of Agriculture (USDA) are pleased to provide this federal interagency report, *The Future of Aquafeeds*. This report was prepared as part of the ongoing NOAA-USDA *Alternative Feeds Initiative* which was launched in 2007.

The purpose of the initiative is to accelerate the development and use of alternative dietary ingredients that will allow the global aquaculture industry to grow without putting unsustainable pressure on industrial fisheries, while maintaining the important human health benefits of diverse aquaculture food products. Although the production of fish meal and fish oil has been relatively constant for decades, supplies of industrial fisheries are limited, and cannot support increased demand from a growing aquaculture industry. Finding alternatives is critical to the long-term sustainable growth of aquaculture in the

---

* This is an edited, reformatted and augmented version of NOAA Technical Memorandum NMFS-F/SPO-124, issued by the National Oceanic and Atmospheric Administration and the U.S. Department of Agriculture, December 2011.

United States and abroad to meet projected increases in consumer demand for safe, high quality farmed aquatic foods.

NOAA, USDA, and other federal agencies contribute vital support for research, development, and the transfer of alternative feeds technology to industry. This report provides a comprehensive perspective on the current state of knowledge and the challenges and opportunities associated with discovery, development, and commercial use of various feed ingredient alternatives. It was prepared by assembling experts from government, academia, private business, non-profit organizations, and other stakeholders in workshops which examined the economic, human health, environmental, and practical implications of various alternative feedstuff options.

The report also summarizes priorities and future directions for feeds manufacturing and includes seven case studies featuring some of the most promising research on alternative feeds being conducted today along with examples of successful alternatives and how they are being used.

The findings, recommendations, and research priorities contained in this report help inform ongoing research and priorities for new research to be supported by NOAA, USDA, and other public and private partners under the joint federal *Initiative*. We hope you find this report informative and useful.

Executive Summary

Background

In 2007, the National Oceanic and Atmospheric Administration (NOAA) and the U.S. Department of Agriculture (USDA) began a detailed and inclusive consultation with independent experts, government researchers, stakeholders, and the general public to gather and distill information on alternative feeds for aquaculture. The driver for this effort was, and continues to be agency and stake-holder interest in speeding up the development and commercialization of viable alternatives to the fish meal and fish oil used in aquaculture. The goal of the NOAA-USDA initiative is to identify and prioritize research to develop feeds that will allow the aquaculture industry to increase production in a sustainable way that does not put additional pressure on limited wild fisheries, that maintains the human health benefits of seafood, and that minimizes negative environmental effects of the use of alternatives. For this development to be realistic, the alternative also has to be economically viable. Thus we considered a triple bottom line in our evaluation of alternatives. These bottom lines take in to account the economic, environmental and human health implications (Figure 1) of alternative feed ingredients.

North America is the worlds largest and most advanced producer of formulated animal diets (followed by the European Union and then China). As a world leader in this area, development and approaches to fish feeds that happen in the United States will help drive change worldwide. It is important to note that even though the US has a relatively small aquaculture sector, developments in aquaculture feeds and advances in technologies and ingredients will have world-wide importance and impact. Currently, the production of feeds for aquaculture worldwide is the most rapidly expanding market in the animal feeds production sector increasing 6-8 percent per year. Aquaculture feeds could represent significant export opportunities for the US feeds sector and their suppliers.

In the United States and worldwide, the development and commercialization of alternative feeds are crucial to the expansion of sustainable finfish and shrimp aquaculture production. Currently, fish meal and fish oil are largely made from small pelagic or reduction fisheries such as anchovies, menhaden, and sardines and from the trimmings of fish processing (both from wild-caught and aquaculture sources). Although the world production of fish meal and fish oil has been relatively constant for the past 20 years, the percentage consumed by aquaculture has risen, now accounting for 60 to 70 percent of the annual production of fish meal and 80 to 90 percent of the annual production of fish oil. Feed for chicken, pork, and pets account for most of the rest, with an increasing percentage of fish oil now going to humans. Pelagic fish are also consumed directly by humans and are used to bait lobster, crab, and fish traps and hooks in commercial and recreational fisheries.

As stocks of pelagic or reduction fisheries used for feed, direct consumption, and bait are limited and already fully utilized, alternate sources of protein and oil are needed for aquaculture feeds. As a potential indication of limited supply, the price of fish meal roughly tripled between 2002 and 2010, and supply remains limited while the demand for fish feed ingredients is expected to continue to rise (Figure 2).

At the same time, prices for farmed salmon and shrimp have been steady or even declined.

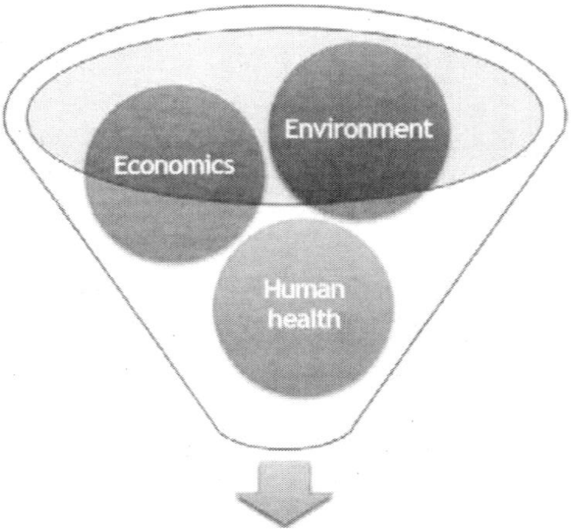

Figure 1. Feeds for Healthy Sustainable Aquaculture.

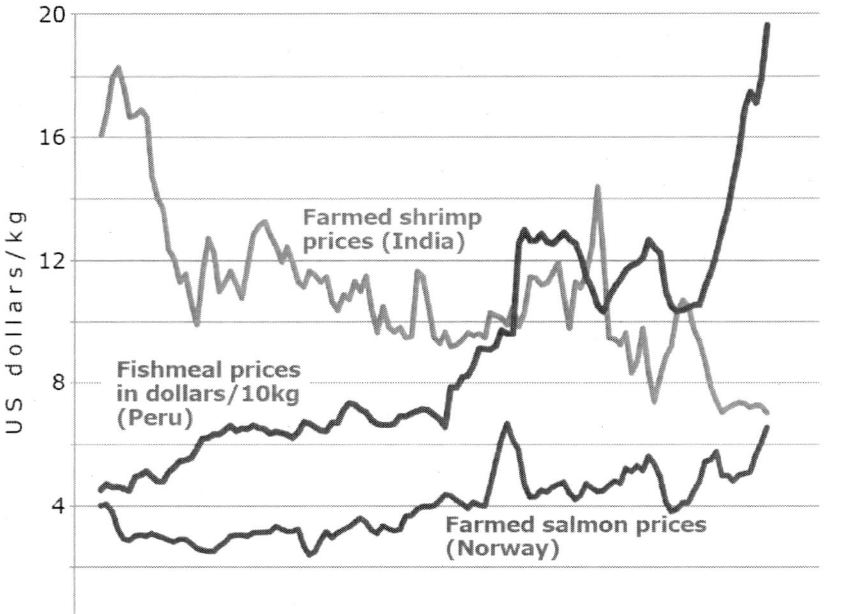

Figure 2. Changes in prices of fishmeal, farmed salmon, and farmed shrimp from 2000-2010.

Environmental considerations may also limit supply. Pelagic fish provide important ecosystem benefits to the marine environment. Although most industrial fisheries are well regulated by catch limits, increased demand for use of forage fish in direct human consumption, for bait, for use in aquaculture and agriculture could provide an incentive to over exploit these fisheries, with negative consequences for the marine environment. Also, changes in fisheries management may further limit supplies of forage fish available. In particular, fisheries managed according to single species sustainable yield measures may not be sustainable from an ecosystem perspective if the importance of forage fish to other animals in the ecosystem is not accounted for. Catch limits or quotas may be reduced to leave a greater supply of forage fish in the oceans to support ecosystem functions.

Developing alternatives to fish meal and fish oil is a global challenge for several reasons. Fish meal and fish oil are worldwide commodities. Asia consumes the majority of fish meal, Europe (especially Norway) is the dominant consumer of fish oil, and South America produces the bulk of both fish meal and fish oil. Fish meal and fish oil are commodities that are traded worldwide. The US is a small player in this market with little control over prices or quantities sold. In addition, the concentrated nature of the product makes supply vulnerable to perturbation, as evidenced by the 2010 earthquake in Chile.

The United States is a small net exporter of fish meal and oil. In 2007 the United States used about 190,000 metric tons of fish meal and 38,250 metric tons of fish oil. Net exports were about 65,500 metric tons of fish meal and 31,000 metric tons of fish oil. Consumption in the United States is mostly for feeds for all types of livestock and pets. A portion of the catch of menhaden, sardines, herring, and anchovies are used for bait for commercial and recreational fishing, fish oil tablets for human consumption, and fertilizer. The majority of fishmeal produced in the United States comes from menhaden, caught in the Gulf and Atlantic followed by meal made from the processing wastes of whitefish caught for human consumption from Alaska. US stocks caught for fish meal and oil production are well regulated under strict management plans mandated by federal law and are not overfished.

This global challenge also represents an opportunity for US agriculture products, seafood processors, and other alternative feed ingredient producers, particularly in supplying Asia where most aquaculture production occurs. The opportunities for US feed and feedstuff suppliers could be significant, and the United States is well poised to take advantage of this opportunity due to our strong agriculture production sector, quality fish nutrition labs, and developed feeds infrastructure.

In November 2007, NOAA and USDA launched the Alternative Feeds Initiative with a solicitation for public comments on several specific questions related to alternative feeds for aquaculture. The questions, which were published in a Federal Register notice included the following:

1. Where should the federal government focus its research efforts in the area of alternative feeds for aquaculture? Are there specific areas that the federal government should not address?
2. What are potential alternative sources of protein and oil for aquaculture feeds? For example, are there specific opportunities for greater use of seafood processing waste and other agricultural by-products in aquaculture feeds? Are there specific obstacles to using these alternatives as alternative dietary ingredients in aquaculture feed?
3. What type of treatments or processes show promise for improvement of existing aquaculture feedstuffs and for developing new feedstuffs? How soon could these technologies be commercialized?
4. Fish meal and fish oil contribute important human nutritional components to aquaculture feeds such as omega-3 fatty acids. If the aquaculture feeds industry seeks to replace fish meal and fish oil with alternatives, how can the nutritional benefits of farmed seafood be maintained or enhanced? For example, what technologies exist for producing omega-3 fatty acids?

Following the initial public comment phase, NOAA and USDA assembled expert panels to address these same four questions and to identify other issues for consideration in the preparation of a rational, fact-based plan to identify and prioritize research and development needs. The initiative's first panel was composed of scientists with expertise in feeds and feed ingredient research, fish and human nutrition, bioenergy, processing, agriculture, and related areas. The second panel was composed of stakeholderers from academia, industry, non-government organizations, and government who had expertise and/or interest in the topic. Government officials with responsibility for research, funding priorities, regulations, and policy observed panel workshops.

In addition to answering the Federal Register questions, panels were asked to identify constraints and concerns about feed ingredients—those currently in use and those that might be used in the future. Panels were also asked to identify possible solutions to the challenge of replacing fish meal and fish oil in future feeds, identify key research and technological challenges associated with developing viable alternate protein and oil sources, and predict the future of feeds for aquaculture—specifically, the challenges and changes that aquaculture will face and the developments that will affect both producers and consumers in next 5 years and in the next 25 years.

A brief summary of panel findings and conclusions follows. Several researchers and other experts were also asked to develop short case studies to highlight specific advances being made in the development of alternative ingredients. Those case studies are included right after the summary of findings.

SUMMARY OF FINDINGS

1. ***Fish meal and fish oil are not nutritionally required for farmed fish to grow.***
   About 40 nutrients—such as essential amino acids, vitamins, minerals, and fatty acids—are required but they can be obtained from sources other than fish meal and fish oil. Fish meal and fish oil have been the preferred ingredients in fish feeds because they contain these nutrients in nearly perfect balance, are easily digestible by the fish, result in good growth and survival, and provide human health benefits. Combining other ingredients to get the same balance is possible, but will require fully understood fish requirements and alternative performance.
2. ***Farming of fish is a very efficient way to produce animal protein and other human nutritional needs.***
   Farmed fish use their feed very efficiently. For example, farmed Atlantic salmon can convert approximately one kilogram of feed (dry) into one kilogram of flesh (wet). In contrast, the feed conversion of poultry is 3-5:1, and pork is 8:1. Fish need fewer calories because they are cold-blooded and they do not need to support their weight.
3. ***Feed manufacturers making diets for carnivorous fish and shrimp have already reduced their reliance on fish meal and fish oil.***
   Application of previous research led to cost-effective substitution using alternatives, which helped mitigate feed costs in the face of increasing fish meal prices (see Figure 2 on page 2). In the past 15 years the ratio of fish in to fish out has dropped from 3-4:1 to

approximately 1.5:1 for major aquaculture species due to increased use of protein and oils in diets from non-marine sources. Fish meal and fish oil are likely to be increasingly reserved for use in specialty diets (broodstock and larval diets) and finishing diets to maintain the human health benefits of farmed seafood.

4. ***Economics is currently the major driver of using alternate feed ingredients in feed mills.***
   Feed producers make substitutions for fish meal and fish oil according to how their price compares with allowable alternatives (i.e., alternatives for which sufficient nutritional and production knowledge and experience exists to allow their use). Panels identified some crucial factors limiting changes to feed formulations, including insufficient information on nutrient requirements of farmed species, especially newly domesticated species, and on available nutrient content and nutritional value of alternative ingredients for fish and shrimp. This area requires investments in research to help feed producers understand the costs and benefits of including alternative ingredients in aquaculture feeds.
5. ***The net environmental effects of the production and use of alternate feeds should be considered.***
   Consideration should be given to the environmental impacts of making dietary changes to feeds for farmed aquatic organisms.
6. ***The human health implications of using alternative feeds needs to be better understood and considered.***
   Long chain omega-3 fatty acids and other nutritional compounds found in fish meal and fish oil provide important human health benefits. Seafood reared on alternative feeds must continue to provide these health benefits to consumers. Human health considerations should be addressed along with economic and environmental considerations when alternatives are considered. To accomplish this, fish nutritionists should work with human nutritionists and food scientists on promising alternative ingredients to determine impacts of alternatives on final product quality.
7. ***Fish meal and fish oil are minor contributors to the world protein and edible oil supply.***
   In 2007, fish meal accounted for approximately 2.3 percent of total protein meals and fish oil for about 2.0 percent of total edible oils. The largest supply of protein on Earth is from soybeans. A 4 percent increase in soy protein meals would nearly equal the total world fish

meal supply. Fish meal and fish oil are likely to continue to be important ingredients, but as supply is limited, they will increasingly be used in combination with other ingredients or for special diets. Substitution will depend on supply, price, and performance of alternatives.

8. ***Recovery and utilization of fisheries processing waste should be encouraged and increased.***

   This material has been shown to produce products of similar biological value to fish meals and oils made from industrial fisheries. The total worldwide amount of fish processing waste from wild capture and aquaculture may equal the amount of forage fish used for fish meal and fish oil from industrial fisheries. But fish processing waste is often not economical to capture because of logistical and technical constraints. Research and financing is needed to help capture the waste products from wild capture fisheries that often are located in remote or inaccessible regions with poor infrastructure. Likewise, research to capture and reuse the waste products from aquaculture should be undertaken. The use of processing waste from aqua-cultured organisms to produce fish meal and fish oil eventually could make aquaculture a net producer of fish meal and oil.

9. ***Plants produce the vast majority of protein and edible oils in the world, accounting for 94 percent of total protein production and 86 percent of total edible oil production.***

   Plants also make up a substantial proportion of diets for carnivorous fish (e.g., 50-60 percent of a typical salmon diet), and that proportion is increasing. It is likely that plants will deliver the bulk of amino acids and fats to diets for farmed fish in the future due to abundance, the potential for increased production, and low cost. Research to increase the use of sustainable plant products in feeds for aquatic organisms will help to increase the importance of agriculture to aquaculture and vice versa. This area of research would be as important to farmers as to aquaculturists and may represent a significant opportunity for American farmers.

10. ***Algae-based biofuel may present opportunities for feed ingredient production because protein is a byproduct of oil recovery from algae, and marine algae produce the long chain omega-3 fatty acids and certain amino acids important to fish and human health.***

    It is too early to understand the ramifications of increased algae biomass production for fish diets, and this area will require communication between algae biofuel scientists and fish nutritionists. Support of research in this area is justified for producing the long chain omega-3 fatty acids alone; a potentially higher value product than biofuel.

11. ***There will likely be increased demand for and production of ethonol and bioplastics. Byproducts from these industries could make good ingredients for fish diets.***
    Fish feeds are mostly made up of protein and oils. Ethanol and some bio-plastic are made from the carbohydrate fraction of plants, leaving behind the protein and oils. Future biofuel production may be quite different from today's focus on ethanol made from corn carbohydrates, which uses a process that degrades the quality of protein waste products. If grain remains a feedstock for ethanol production, new approaches to recover high-quality protein and oil from the ethanol production process will be needed to make it suitable for wide spread use in fish feeds. Biodiesel is made from the oil fraction, leaving behind concentrated protein that is already suitable for fish. Fish nutrition researchers should work, and coordinate with, biofuel scientists to ensure byproducts are safe and usable for fish. Research that supports processes resulting in high-quality protein and oil byproducts of fuels production should be encouraged.
12. ***As replacements, many alternatives are higher in cost per unit fish gain (biological value) than fish meal and fish oil.***
    However, the recent trend (since 2006) has been for fish meal and fish oil prices to increase faster than prices of alternative protein and oil sources. Research that can help lower costs or improve the biological value, without raising costs, will increase the rate of fish meal and fish oil replacement.
13. ***Fish have dietary needs and preferences for specific compounds not found in plants, so there is a need for specialized products that supply these compounds and/ or add flavor to the diet.***
    These ingredients will likely be higher in cost than the bulk protein and oil products and will need to contain flavors, nutrients, or properties not found in bulk proteins and oils but which are needed for fast growth, health or increase consumption. Examples are algae, invertebrates, animal by-products and seafood trimming meals and oils. Additional ingredients such immune system enhancers are also beneficial to enable use of higher levels of alternatives. Research is

needed to develop materials that will enable greater use of cheaper more abundant protein meals and oils.

14. ***Alternative sources of protein and oil are common commodities used in livestock and companion animal feeds and come from novel byproducts from other industries, underutilized resources, or completely novel products.***
    - Existing commodities that have the potential for greater use in feeds include protein concentrates from grains or oilseeds and byproducts from animal proteins.
    - Novel byproducts from other industries include proteins recovered from biofuel production or single-cell proteins produced from inexpensive carbon sources.
    - Other sources include fish processing wastes, trimmings and/or bycatch from fishing.
    - New products including meals produced from worms, insects, and marine invertebrates, and meals and oils from algae.

    What these products have in common is that they are underused and/or underdeveloped protein and oil sources that require variable degrees of investment in research and development to become more widely used. Some possess attributes that are detrimental to fish (e.g., anti-nutrients), or they contain insufficient levels of essential or semi-essential nutrients and need to be processed, blended with complementary products or supplemented. More information is also needed to evaluate the environmental impacts associated with using various feed ingredients. Information on contaminant content of alternate products is also needed to place risks and benefits to fish wellness and human health into a rational context. Coupled with this is the opportunity to maintain or improve the safety and healthfulness of farmed fish products for the consumer by using alternate ingredients. All these topics will require investments in research and development.

15. ***Plants and other alternatives contain some compounds (anti-nutrients) that are detrimental to fish.***

    Although there are processes to remove or inactivate many of these compounds, further research and development is necessary to improve these processes. Fish may also be selectively bred to be relatively more tolerant of the anti-nutrients in some alternatives.

16. ***Harvest of lower trophic level species, such as krill, for fish meal and oil production may be possible, but the environmental benefits afforded to the marine ecosystem from these species should be considered along with the economic and nutritional aspects of their use.***
    While this may provide an option in the near term, the harvest of any wild population, including krill, would require careful management and would be limited to what nature can supply.
17. ***The use of bycatch for production of fish meal and fish oil could provide a substantial amount of these products without increasing the current impact from the wild capture fisheries.***
    Although traditional processes exist to convert bycatch into fish meal and fish oil, concerns over creating a market for non-target species and the logistical issues associated with dealing with retained bycatch at sea have been expressed.
18. ***Demand for long chain omega-3 fatty acids for both direct human consumption and feed ingredients is likely to increase beyond the amounts available from marine resources.***
    Alternative sources are needed and should be developed, such as algae, microorganisms, and/or oilseeds. More efficient use of long chain omega-3 fatty acids can be made in aquaculture through improvements in feeding practices and formulation. Research leading to new cost-effective sources of long chain omega-3 fatty acids will benefit human health as well. Research to improve production and the efficiency of use should also be supported.
19. ***Farmed fish species are being increasingly domesticated and performance is improving through conventional genetic selection and selection for performance on plant-based and /or low fish meal based aquafeeds.***
    As aquatic species are domesticated, selection can be directed toward better use of non-fish meal and non-fish-oil ingredients.
20. ***Scientific information on the nutritional requirements of farmed fish species, and feed ingredients, and the interaction between the fish and the diet, will need to expand greatly to make substantial improvements in feed formulation by commercial aquaculture feed producers.***
    Updating the National Research Council (NRC) requirements for fish on a regular basis and support for research that helps define the basic

nutritional requirements for farmed aquatic species should be supported.

Case Study One: Developing the Potential of Fish Processing by Products Takes Guts

Dr. Peter Bechtel, Pete Nicklason, Dr. Michael Rust and Dr. Scott Smiley

If you ask a pork producer about how much of his product is actually used by society, he is likely to reply, "We use everything but the squeal". In fact, much of agriculture has long made good use of the various parts of animals that humans don't eat. Much of these so called 'byproducts' or 'co-products' are usually processed into feed for pets and livestock. While the recovery and repurposing of all parts of terrestrial animals is quite efficient, recovery of fish parts is just starting to improve.

If you buy a nice one pound fillet at your local fishmonger, it is likely that about one pound of other potentially useful stuff has also been generated in cutting that fillet. Various terms such as 'fish trimmings', 'fish wastes', 'fisheries byproducts', 'fisheries coproducts', 'fish scraps', and even 'fish offal' (pronounced "awful") are used to describe various components of the heads, guts, fins, bones, and skin that are left over after cutting two fillets from a whole intact fish.

Nowhere is the potential to effectively capture this valuable material greater than in the State of Alaska. The Alaskan seafood industry harvests more than half of the total U.S. commercial fish catch each year and processing this harvest into food for people leaves over 1.1 million tons of fish processing waste. Experts have estimated that a quarter of this waste may be discarded and it's potential value lost.

The challenge in Alaska and elsewhere, is that much of this fish processing waste is created seasonally in remote areas with poor infrastructure. Given the biology of the fish, they are often available in huge amounts but for only a very short season. Because fish spoil so quickly and because a huge volume of fish processing waste is produced in such a short period of time, it requires big expensive equipment and a good deal of energy to process it all before it spoils. After the season, there is nothing to do with that expensive equipment until the next harvest occurs. For some fisheries this might mean a six week-long season followed by the rest of the year off.

Aside from the huge processing operations along the shore of the Bering Sea, small processors specializing in remote and seasonal fisheries, such as salmon, rarely generate sufficient volumes of fish processing waste to justify investment in large scale equipment. The intermittent nature of the fishery, and the remote seasonal nature of the locations the fish are processed, mean that traditional approaches such as those used by the year round pork and chicken industries are not cost-effective when it comes to fish, so waste material ends up back in the sea.

Driven by the sheer potential for economic and environmental gains, state and federal researchers are developing techniques that will help industry and local communities tap into the value of fish processing waste from Alaska and elsewhere. For some of the larger fisheries that harvest over a longer period of time, a great deal of progress has been made with modifications to traditional solutions. For example most of the processing byproducts from pollock (one of the world's largest human food fisheries) and other white fish are used to make fish meal and oil. This has already made Alaska the second largest producer of fish meal in the United States. But the pollock fishery is huge with roughly 1 million metric tons harvested annually making a more traditional processing approach feasible with relatively small modifications. Additionally, government regulations required processors building new seafood processing plants along the Bering Sea to include machinery designed to effectively handle seafood processing byproducts.

An important and more difficult challenge facing researchers remains to develop reliable methods to stabilize small volume fishery processing waste until it can be dried and worked on after the initial hectic processing season closes. This is important because many Alaskan salmon processors are small, seasonal and remote so they do not have fish meal machinery necessary to handle their byproducts before they spoil.

Working collaboratively, researchers have shown that the byproducts of seafood processing can be converted into commercially valuable products including protein meals, oils, and more. In addition, researchers are finding that diets incorporating protein from fish processing waste and used for feeding fish, pigs, poultry, dogs, cats, and even reindeer, are equal in nutritional value to those made with traditional fish meal and oil. Significant progress has also been made by supporting research on the development of new cost-effective processing methods as well as the development of new feed ingredients.

Several collaborative studies demonstrated that seafood byproducts could be stabilized against microbial degradation for the short term (weeks to months) by lowering the pH through the addition of acids–a process similar to how yogurt is stabilized. This stabilization allowed for room temperature storage. Most important, the resulting dried meal remained suitable as a protein feed ingredient for salmon and trout diets. Recently, other stabilization strategies show promise as well, such as using both lactic acid bacterial fermentation and chemical acidification on Alaskan seafood processing byproducts.

Even higher value products can be made from some of the components in the seafood waste stream. Further separation of the components of seafood processing waste can be used to create higher value specialty protein and oil products that may offer a solution to supplementing nutrient deficiencies in plant protein meals and oils being developed for use in aqua-culture feed formulations. Studies in collaboration with the processing industries have involved the development of tailored protein powders, and the recovery of usable proteins from a variety of seafood processing waste streams. The human health benefits of fish oils from cold-water species, including the long chain poly-unsaturated omega-3 fatty acids, are a good example of a high-value human nutrient that has been developed from this resource.

Photo courtesy AFSC.

Walleye pollock (Theragra chacogramma).

Continued research in this area promises to reduce the seafood industry's environmental impact and increase the economic viability of both the industry and coastal communities. In fact, this research is now being transferred to the industry. At least one new plant is in the design phase based on the new technologies which will recover 17 million pounds of waste a year that is currently going into the sea. Other companies are increasing the extraction of fish oils from their processing byproducts.

Soon, when you ask a fisherman or a fish farmer about how much of his product is actually used by society, he will be able to reply, "We use everything but the gulp!" It turns out that offal doesn't have to be awful after all.

Case Study Two: From Fish Meal-Dependent to Fish Meal-Free: Feeds Research is Producing the Alternative Diets of the Future for Trout

Dr. Fredric Barrows and Dr. Jeffrey Silverstein

For more than 100 years, rainbow trout have been farmed–or 'cultured'–in the United States. Trout are now farmed in all 50 states and are one of our most valuable domestically grown fisheries products. In 2007, US farmers produced more than 57 million pounds of trout for consumers; state and federal hatcheries release more than 147 million trout annually for restoration, conservation, and recreational purposes.

Feeding these hungry trout presents a challenging research opportunity. For the last 50 years, fish meal has been the primary protein source in trout feeds, with levels up to 50 percent being common. With the rising cost of–and increasing global demand for fish meal and fish oil–fish feed manufacturers are under pressure to find alternative ingredients to replace portions of fish meal and fish oil in aquaculture feeds.

So far, research using alternative feed ingredients demonstrates that fish meal is not a required dietary component for trout. Research also indicates that protein ingredients from sustainable sources can replace most of the fish meal in trout feeds as long as care is taken to ensure that all essential dietary nutrients are present in required amounts and in bioavailable forms. As research studies progress, the trout aquaculture industry will be able to move from fish meal-dependent to fish meal-free feeds.

However, it is not yet an easy or economically feasible task to remove or substantially reduce fish meal and fish oil in aquaculture feeds without affecting the growth and health of the fish. U.S. scientists from the disciplines of grain genetics, grain processing, fish nutrition, and fish genetics have teamed up to approach the problem from different but complementary directions. This multidisciplinary approach is producing results that have enabled scientists to formulate and test novel trout feeds that are cost-effective and use significantly less fish meal and fish oil than conventional feeds. These studies are helping feed producers move toward the goal of developing fish meal-free feeds.

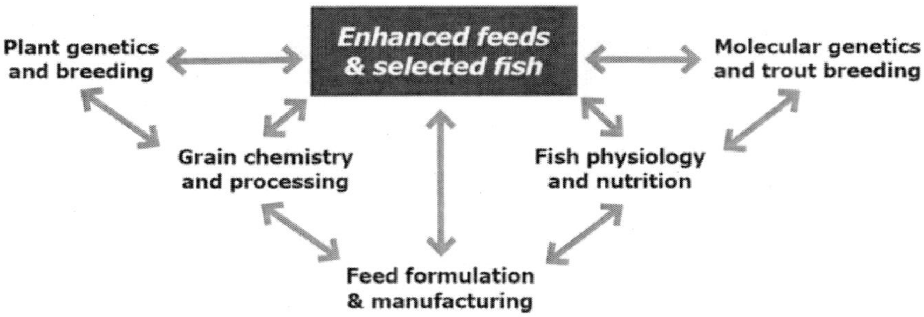

Alternative trout feeds developed to date are considered prototypes, since they are slightly more expensive than conventional feeds. Further, growth rates of trout remain about 10 percent slower than when a fish meal-based feed is used. However, the new feeds are in the early stages of development and the research is beginning to show that total elimination of fish meal from finfish diets is not only feasible, it also has other performance benefits.

For example, research is showing that supplementation of specific nutrients to a plant-based fish diet–including the amino acid, taurine, as well as electrolytes, and higher levels of specific vitamins–results in weight gains equal to trout fed fish meal-based feeds. The resulting fillets are also just as flavorful and nutritious as the fillets from trout fed fish mealbased diets.

Case Study Three: Plant-Based Feeds for Black Seabass Show Promise

Dr. Wade Watanabe and Dr. Md Shah Alam

Scientists at the University of North Carolina Wilmington-Center for Marine Science (UNCW-CMS) are developing alternative plant protein based practical diets for the culture of black sea bass, *Centropristis striata*, a commercially important species found in waters along the Atlantic coast from the Gulf of Maine to northern Florida. Their wide acceptance as an excellent food fish and their high market value has led to over-harvesting of wild stocks especially in the South Atlantic U.S. coast. Increased awareness of the status of the black sea bass populations, coupled with high market value and demand, has led to an interest in the development of culture technologies for commercial production. Reliable protocols for spawning and larval rearing of black sea bass are already established. At UNCW, a team of researchers lead by Dr. Wade O. Watanabe, is developing nursery and grow-out technologies for producing marketable fish in recirculating aquaculture systems. Nutrition and diet development are critical components of their research.

Black sea bass grow rapidly when fed prepared feeds consisting largely of marine feedstuffs such as menhaden fish meal or natural foods such as live tilapia. The future potential for limited availability and rising cost of fish meal and live tilapia limit their application in future practical diets.

Soybean meal is considered to be one of the most suitable and stable supplies of an alternative ingredient for replacing fish meal in commercial fish feeds. Compared to other grains and oilseeds, soybeans are promising because of their high protein content high digestibility and good amino acid profile. A series of experiments were conducted by UNCW fish nutritionist Dr. Md Shah Alam with the assistance of graduate student Katherine Sullivan to test the effect of varying dietary levels of solvent-extracted soybean meal supplemented with or without amino acids and attractants in the diets of juvenile black sea bass. Diets were formulated to replace menhaden fish meal protein with solvent-extracted soybean meal protein at 0 to 100 percent with or without supplementing amino acids and 1 percent attractants (taurine, betaine, glycine and alanine). All diets were formulated to have about the same crude protein and same oil level for each experiment. To enhance palatability, all diets contained 5 percent krill meal and 7.5 percent squid meal. Results of these experiments showed that the maximum level of menhaden fish meal protein replacement with solvent extracted soybean meal protein was 70 percent with 1 percent attractants, 7.5 percent squid meal and 5 percent krill meal and with or without supplementing methionine and lysine in the diets. Greater than 70 percent replacement of fish meal protein with soybean meal caused growth, whole body protein and oil to decrease. Similar trends were observed for feed efficiency, specific growth rate, feed intake and protein efficiency ratio. These short-term laboratory based studies were extended to pilot-scale grow-out conditions. An experiment to test the replacement of fish meal protein by soybean meal protein without adding squid meal, krill meal and attractants is currently in progress. UNCW scientists are also conducting research to test the flavor and nutritional value of the fish fed the high level of soybean based diets and comparison with the fish fed fish meal-based diets. Results to date indicated that black sea bass exhibited excellent growth when fed feeds containing relatively high levels of soybean meal. These results will be used to develop environmentally-sound and cost-effective plant protein-based feeds for black sea bass aquaculture.

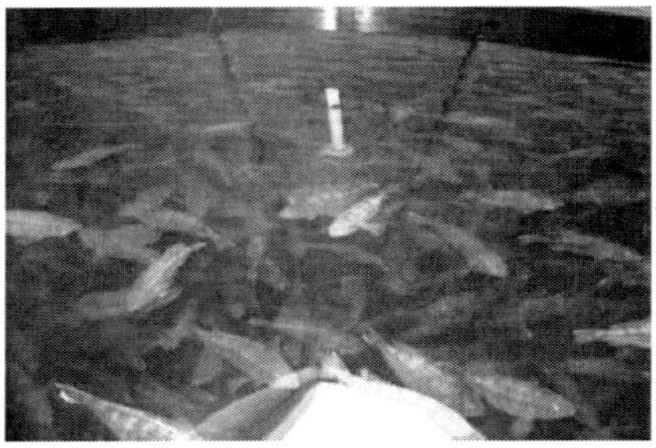

Hatchery raised black sea bass grown to market size in a recirculating tank system at the University of North Carolina Wilmington, Center for Marine Science.

Experimental feeds formulated and produced at the University of North Carolina Wilmington, Center for Marine Science.

CASE STUDY FOUR: SHRIMP FARMERS JOIN WITH RESEARCHERS TO TEST BEST NEW DIETS

Dr. D. Allen Davis

The values of shrimp seafood products are declining, stable or in some cases increasing only slightly which is wonderful news for the consumer. At the same time feed, fuel and processing costs are rapidly increasing and causing some U.S. commercial aquaculture operations to cut production costs or go out of business. Given a fixed formulation, the cost of shrimp feeds has almost doubled in the last two years. This is in response to a wide number of factors but was first triggered by rising and unprecedented increase in the price of fish meal from 2006 to 2008. The rapid increase in world fish meal prices was followed by a moderate increase in the cost of other protein sources and a recent rapid rise in grain prices. All of these taken together caused feed costs to nearly double without corresponding increases in the value of the final product.

Although fish nutrition research cannot change world prices, it can provide alternative formulations to moderate feed price increases. Fortunately for the shrimp industry, researchers have been working towards the goal of quantifying nutrient requirements and providing information on the use of alternative feed ingredients for some time. Traditional shrimp feed formulations include 20 to 30% fish meal which is one of the most costly protein sources. Fish meal is an ingredient for which world supply cannot be expanded and is considered a limiting factor for the continued expansion of aquaculture. At this point, experts agree that one of the first steps to reduce shrimp feed prices is to provide alternatives to the use of marine ingredients and develop plant-based diets.

A number of research groups have been systematically identifying nutrient requirements and working to improve feed formulation technologies. Using a variety of funding sources, researchers at Auburn University, Texas A&M, and the Wadell Mariculture Center took the challenge to develop a synergistic program to identify limiting nutrients in plant-based diets and demonstrate their findings to the industry. This included a systematic approach to identifying limiting nutrients in the laboratory, testing diet formulations in outdoor tanks and then in research ponds. Using balanced formulations based on alternative protein sources, primarily of plant origin, resulted in an improvement in the overall nutritional quality of practical diet formulations as well as considerable reductions in formulations costs. These formulations were made possible by systematically identifying limiting nutrients and balancing the formulations as fish meal was removed. For shrimp, identifying the total sulfur amino acid requirement was a key factor to removal of fish meal. Once the fish meal was removed, researchers also discovered essential fatty acids, cholesterol and phosphorus needed to be supplemented to plant-based feeds. Numerous studies have been conducted which have demonstrated the feasibility of reducing or completely replacing fish meal with no adverse effect on the productive performance of *L. vannamei*. These and current trials have demonstrated that practical diets can be formulated using soybean meal as the primary protein source. Other renewable protein sources such as distiller's grain solubles, pea meal and corn gluten meal have been utilized in combination with soybean meal to enhance the amino acid balance and to diversify the ingredient base of these formulations.

Based on these results, the next step was to gain commercial acceptance of reduced fish meal formulations. In order to promote the continued development of quality feed formulations, the American Soybean Association–International Marketing provided funds to transfer feed formulation technologies to the shrimp industry. The project involved working directly with feed mill manufactures and producers in the United States, Latin America (Ecuador, Mexico, Colombia, Venezuela) and more recently Asia to provide technical assistance to both feed mills and shrimp farmers allowing them to improve production practices. With this purpose in mind, a series of regional seminars was conducted in each country to disseminate results obtained at various research centers and to provide sound technical advice on feed manufacturing and culture technologies. Producers and feed manufacturers who were willing to try new practices and improve their operations using plant-based feeds were identified. Shrimp farmers were asked to compare their conventional fish meal-based feed with feed containing less fish meal. The level of fish meal reduction was determined by the producers and feed manufacturer to make them feel comfortable with the experiment. Although controlling experimental conditions in the real world is difficult, farmers were asked to follow a protocol that included using similar sized ponds, using shrimp larvae of the same origin and stocking at similar densities. The production protocols, particularly feed inputs, were reviewed by experts to provide additional technical support to further encourage improved management practices. A similar approach was used with the feed mill manufacturers. Technical support was provided in terms of reviewing manufacturing practices, feed formulation and feed management recommendations. In most cases, improvements in feed manufacturing processes, formulation restrictions and feeding tables were made, resulting in significant reductions in manufacturing costs. Results to date have been encouraging with most farmers and feed mill manufactures adopting the suggested improvements in feed formulations technologies.

CASE STUDY FIVE: SEAWEED FARMING MAY BE KEY FOR ALTERNATIVE AQUACULTURE FEEDS

Dr. John Forster

Could seaweeds be used in aquaculture feeds? The answer appears to be yes.

While terrestrial ingredients have served as the immediate replacements for fish meal in aquaculture feeds, the possible use of seaweeds in the future holds the promise that marine aquaculture could, one day, be sustained without using land, fresh water, or fertilizer.

Presently, there are about 14 million metric tons (15.4 million US tons) of seaweed farmed worldwide compared to about one mmt harvested from the wild. Most seaweed is farmed in Asia where about half is used for human consumption and half for industrial processing into marine colloids (used to make paint and foods smooth) and other products. Extrapolation from Chinese production of a seaweed known as *Laminaria*, for which the average yield in 2004 was 19.4 mt dry weight per hectare (8.7 US tons per acre) suggests that the world's present total agricultural output of about five billion mt (5.5 billion US tons) could be matched by seaweed production in an area of 2.6 million km^2 (1.0 million square miles). This is only 0.74 percent of the 349 million km^2 (135 million square miles) covered by the world's oceans.

Of course extrapolations can mislead and any development on this scale will take decades, if not centuries. But the potential and its significance are obvious and they prompt a second question: should this be a priority for future research? Again, the answer appears to be yes.

Seaweeds contain valuable amino acids and fatty acids, including omega-3 and omega-6 polyunsaturated fatty acids, marine oils which are part of the reason why seafood is considered healthful–see the fatty acid table. They also contain minerals, vitamins and anti-oxidants as well as being rich in dietary fiber but low in starchy carbohydrates, thereby providing less glycemic load than ingredients derived from grains.

However, several impediments to utilization in feeds must be overcome before seaweed products can be used, including:

1. **Processing (bio-refining) of seaweed to increase the availability of nutrients**
   In the raw state, seaweed nutrients are protected by indigestible cell walls, or are chemically bound in a way that diminishes their potential nutritional value. Japanese scientists have used fermentation and enzymatic digestion to release and concentrate these nutrients. They have shown that when protoplasts, which contain the bulk of the protein, from the seaweed *Porphyra* are released from the cells after treatment with a polysaccharide degrading enzyme, their inclusion at only 5 percent in feeds for red sea bream led to improved growth and nutrient retention. Significantly, *Porphyra* is known to contain high levels of taurine, an amino acid that has also been found to stimulate growth of fish when they are fed with land plant-based feeds.
2. **Species selection and breeding of seaweeds for fast growth and high concentrations of the most valuable nutrients**
   The science of seaweed breeding lags well behind that of terrestrial plants. However, in China, where most of the world's seaweed is farmed, one variety of *Laminaria* has undergone 15 generations of selection from 1959 to 1974, and grows faster and contains more iodine compared with the natural population. Advances have also been made in protoplast separation and fusion that can produce seaweeds with improved traits through recombination of naturally occurring genes. Northeastern University in Boston is a leader in this field having developed a strain of 'super nori'.
3. **Development of large-scale, deep water farming methods to increase the ocean area that can be used for seaweed production**
   Previous U.S. research from 1968–1990, contemplated large-scale seaweed farming for biomethane production. A new initiative in Japan, similarly contemplates large farms for bio-ethanol production. In both cases, byproducts from energy production might be suitable for aquaculture feeds. But the challenges of large-scale farming for bioenergy are daunting. Perhaps, if research focused on the value of seaweeds as feed, it could start on a scale that is more practical and requires less investment.

Type of fatty acid (percent of total fatty acid content)

Seaweed	Saturated	Monosaturated	PUFA's	Ω6 PUFA's	Ω3 PUFA's	Ω6/Ω3 Ratio
Himanthalia elongate	39.06	22.75	38.16	15.08	18.70	0.81
Laminaria ochraleuca	33.82	19.23	46.94	20.09	25.08	0.83
Undaria pinnatifida	20.39	10.50	69.11	22.10	44.70	0.49
Palmaria sp.	60.48	10.67	28.86	2.14	25.52	0.13
Porphyra sp.	64.95	18.01	16.10	7.97	7.20	1.21

From MacArtain et al., 2007.

In the western hemisphere modern aquaculture focuses on the production of aquatic animals. But as seen in terrestrial agriculture, plants support the needs for food, feed and other goods. Is it unreasonable to suppose that, one day, aquatic plants might be able to do the same?

Case Study Six: Research on Diets for Threatened and Endangered Fish Species Held in Captivity Gains Ground

Dr. Ronald G. Twibell and Dr. Ann L. Gannam

For more than a century, the National Fish Hatchery System of the US Fish and Wildlife Service (FWS) has produced economically important species, such as Chinook salmon and rainbow trout, for mitigation purposes. Currently, more than 100 species of plants and animals are reared at the Service's 70 National Fish Hatcheries (NFHs) and seven Fish Technology Centers (FTCs). More than 1.9 million pounds of rainbow trout alone are produced annually by the Service. Considering the amount of fish feed used at NFHs and FTCs each year, the Service is a significant commercial fish feed consumer.

Courtesy of USFWS.

Lahontan cutthroat trout.

Courtesy of USFWS.

Bull trout.

Numerous federally listed species are reared at the agency's hatcheries and technology centers, to support the recovery plans for many aquatic species which recommend propagation of captive fish. The objective of these captive rearing programs generally is to augment remaining natural populations until habitat conditions improve. A few examples of federally protected species currently reared by the Service include bonytail, bull trout, fountain darter, pallid sturgeon, razorback sucker, Lahontan cutthroat trout, and Rio Grande silvery minnow. Among other challenges, hatchery personnel must identify appropriate commercial feed for those and other species reared in captivity.

Aquaculture feeds used in the US have been developed to meet the specific nutritional requirements of a few species, including channel catfish, rainbow trout, salmon and tilapia. The commercial diets available for those species are the result of decades of research conducted by fish nutritionists and feed companies. Even today, fish nutritionists continue to improve diets for commercially important species. But, limited research has been conducted on the basic nutritional needs of any of the federally listed threatened or endangered species of fish reared in captivity.

Federally protected species of fish exhibit feeding behaviors ranging from strictly carnivorous to strictly herbivorous. Other than general information on feeding behavior and diet in the wild, relatively little is known about the nutritional needs of these species. Furthermore, some protected species may complete their entire life cycle at a national hatchery and have access only to commercial fish food. As commercial feeds evolve and more alternative ingredients are included in the formulations they will have to meet the nutritional requirements for the entire life cycle of those species and have no detrimental effects on their health or reproduction over long periods of time. Currently, Service personnel are attempting to identify the most appropriate commercial diets for various protected species. For example, studies evaluating commercial diets for Rio Grande silvery minnow, bonytail, and Atlantic sturgeon were published recently. Research on nutritional needs of protected aquatic species is ongoing and is essential for the successful propagation of these animals.

Photo courtesy USFWS.

Production of fish feed at Abernathy Fish Technology Center.

CASE STUDY SEVEN: SOY PRODUCTS AND AQUACULTURE ARE A WINNING COMBINATION

Dr. Michael Cremer

A market for over six million metric tons of soybean meal has been successfully created over the past 15 years through the development, field testing, and demonstration of all-plant protein, soymeal-based feeds to fish farmers in China. Opening this market to alternative feeds has helped boost China's freshwater aquaculture production from less than five million metric tons to more than 20 million metric tons (5.5 to 22.0 million US tons) by alleviating the necessity for traditional animal protein sources, such as fish meal, in most freshwater fish diets. In the process, it has helped the Chinese aquaculture industry advance from traditional manure-fertilized to modern, feed-based production of the majority of carp, tilapia, catfish and other freshwater fish species farmed in China. This new approach to feed has provided domestic and international consumers with ready access to higher quality farmed seafood from China at reasonable prices, while providing a growing market for US grown soybean products.

Soy products can make up 50 percent or more of the feeds for the carp and tilapia species that comprise nearly two-thirds of the freshwater aquaculture production in China. As an example, a recent pond feeding trial conducted by the American Soybean Association International Marketing program from 2006 through 2008, demonstrated that a 60 percent soy product, all-plant protein feed for grass carp yielded up to 65 percent higher production and up to 500 percent greater profit when compared head to head with a traditional Chinese poly-culture system that used a combination of feed and grass. A 55 percent soy diet for tilapia not only grows tilapia quickly and with a high feed conversion efficiency, but it provides a healthy 2:1 ratio of omega 6 to omega 3 fatty acids. Other studies have demonstrated that the typical 20 percent fish meal inclusion in fingerling feeds for carp and tilapia can be fully replaced with soy protein concentrate, further alleviating the demand on limited fish meal stocks. The soy-based feeds additionally blend soy and fish oils to reduce dependence on fish oil stocks.

Marine fish and shrimp producers worldwide are also gaining knowledge from research conducted by the US soybean industry. Currently, research is underway to boost soy product inclusion in the diets of key marine fish and shrimp cultured in Asia, Europe, the Middle East, Latin America, and the U.S.

A recent study in which soybean meal and soy protein concentrate replaced all but 10 percent of the fish meal in a test diet was successfully demonstrated with pompano farmed in open ocean cages in southern China. Studies in Spain have demonstrated that the protein contribution from fish meal can be reduced to as low as 15 percent in the diet of gilthead sea bream with properly formulated soy feeds, and to 40 percent with European sea bass. A high omega-3 fatty acid soy oil is being tested as a fish oil replacement for yellowtail cultured in open ocean cages in Hawaii. New soy-based diets for white shrimp also have been developed and are being demonstrated throughout Latin America and Asia.

Collectively these studies are reducing the requirement for fish meal and fish oil in aquaculture feeds and helping to reduce aquaculture's environmental impact, and increase industry sustainability. For additional information on soy use in aquaculture, see www.soyaqua.org.

A three-year pond feeding trial (2006–2008) in China demonstrated that a 60 percent soy product, all-plant protein feed for grass carp yielded up to 65 percent higher production and up to 500 percent greater profit when compared head-to-head with a traditional Chinese polyculture system that used a combination of feed and grass, while simultaneously reducing the environmental impact by 50 percent or more.

Pompano *Trachinotus blochii* grew from 25g to 610g in 146 days on a diet in which soybean meal and soy protein concentrate were the primary protein sources, and in which fishmeal inclusion was reduced to 10 percent of the diet (ASA-IM feeding trial, Hainan Island, China 2007).

Pangasius catfish grew from 0.1g to 880g in 181 days on soy-based, all-plant protein feeds with an feed conversion ratio (FCR) of 1.2kg feed to 1.0kg fish gain.

Approach & Processes

To meet growing consumer demand for seafood in the United States and world-wide, increasing supplies of finfish, shellfish and other seafood products will be needed. Most experts agree that development of aquaculture will be needed to meet this increase in demand. The challenge is how to ensure that increases in aquaculture production are sustainable.

The development and expansion of farming of carnivorous fish and shrimp species may soon be constrained by a limited supply of fish meal and fish oil for feeds. Fish meal and fish oil traditionally have made up a large part of the diet of farm-raised carnivorous fish and shrimp. The composition of these feed ingredients is almost perfectly matched to the dietary requirements of fishes. However, there is no dietary requirement for fish meal or fish oil for any organism. Because dietary requirements are not for the ingredients per se but for the nutrients they contain (e.g., amino acids, fatty acids, vitamins, and minerals), feeds that lessen the reliance on these limited ingredients— such as alternative protein and oil sources–can be developed.

For these reasons, the US Department of Commerce's National Oceanic and Atmospheric Administration (NOAA) and the US Department of Agriculture (USDA) sponsored expert and public consultations on the future of aquaculture feeds and the benefits to the U.S. economy by the development of such alternative feeds. These agencies were greatly aided in this effort by help and advice from scientists and others within the Department of the Interior (DOI) and the Food and Drug Administration (FDA).

The Consultation Process

The consultation process consisted of nine parts:

1. An invitation to the general public to comment on the issue, and respond to four questions designed to elicit input of a broad array of suggestions and approaches in the Federal Register. The public comments also helped indicate the level of understanding and knowledge that the public has regarding fish feeds.
2. A consultation with experts who are active researchers in the area of fish feeds, feedstuffs, nutrition, and related topics (scientific experts panel).
3. A consultation with experts who are active stakeholders in the area of fish feeds, feedstuffs, nutrition, and related topics (stakeholder experts panel).
4. Reporting case studies where the shift from reduction fishery fish meal and fish oil to alternatives is already happening or areas that might hold promise for the future. Case studies are featured after the report summary.
5. Futurecasts focused on fish feeds by the attendees at the two experts meetings.

6. Information from parts 1 through 5 above was summarized in a draft of this technical report addressing the questions raised by the public comment process, summarizing the results of the two experts panels, the future casts and reporting on the case studies.
7. A public review of the draft report to provide input before it is finalized.
8. Publication of this final report and outreach to interested parties.

Steering Committee

A steering committee made up of scientists, federal policymakers, and communications experts was assembled to move the process through these steps. These individuals are listed on page iv.

The purpose of the steering committee was to fine-tune the objectives and questions asked, suggest and contact the appropriate scientists, develop dates and locations for panel meetings, choose facilitators for panel meetings, serve as reviewers of the draft report, and develop presentations to be given at public meetings. The editorial subcommittee assembled, wrote portions of, and edited the report. Numerous authors contributed case studies in their areas of expertise and/or in futurecasts.

In conducting this initiative, the steering committee was guided by several principals when considering an ingredient, process, or approach to reduce the use of conventional fish meal and fish oil in aquaculture:

- All ideas were welcome. The committee was more interested in what to do, rather than what not to do. Thus all ideas were welcomed equally. Conversely, objections to various feedstuffs are recorded but were not viewed as justification for their exclusion from consideration or exclusion from the final report.
- The committee attempted to adopt a triple bottom line approach when evaluating alternatives. This meant trying to account for:
    1. The economic performance of an ingredient, process, or approach.
    2. The environmental performance of procuring and using an ingredient, employing a process, or following an approach.
    3. The human health performance of the product resulting from the substitution of an ingredient, process, or approach.

Initial Public Comment Process

The Federal Register notice containing questions to solicit input was published November 16, 2007. The public comment period ended February 29, 2008. The questions from the Federal Register notice were as follows:

- Where should the federal government focus its research efforts in the area of alternative feeds for aquaculture? Are there specific areas that the federal government should not address?
- What are potential alternative sources of protein and oil for aquaculture feeds? For example, are there specific opportunities for greater use of seafood processing waste and other agricultural by-products in aquaculture feeds? Are there specific obstacles to using these alternatives as alternative dietary ingredients in aquaculture feed?
- What type of treatments or processes show promise for improvement of existing aquaculture feedstuffs and for developing new feedstuffs? How soon could these technologies be commercialized?
- Fish meal and fish oil contribute important human nutritional components to aquaculture feeds such as omega-3 fatty acids. As the aquaculture feeds industry seeks to replace fish meal and fish oil with alternatives, how can the nutritional benefits of farmed seafood be maintained or enhanced? For example, what technologies exist for producing omega-3 fatty acids?

Experts Panels

Two groups of experts were assembled to advance a path to the development of commercial aquaculture diets that have no or limited usage of fish meal or oil derived from commercial reduction fisheries. The first panel (scientific experts panel) was made up of scientists actively working and published in feeds and feedstuffs research, fish and human nutrition, bioenergy, processing, agriculture, and related areas. These scientists provided the scientifically defensible options for further development of alternative feed ingredients for aquaculture. This panel was primarily made up of university and government scientists from around the world. Scientists came from Australia, Canada, Japan, Norway and the United States. This panel met at NOAA's Manchester Lab in Washington State in February 2008.

The second panel (stakeholder experts panel) was made up of stakeholders who are experts in the practical areas of feeds, human nutrition, and specific feedstuffs; members of environmental groups and consumer groups; public hatchery system managers and members of the commercial aquaculture industry; and others with expertise related to the topic. They addressed the same charge as the scientific panel. This panel met at NOAA headquarters in Silver Spring, Maryland, in April 2008.

The expert panel workshops were used to capture expert opinions, develop consensus on key issues where possible, and vet options. Observers included government officials who are responsible for setting research funding priorities, regulators, and policymakers at the agencies with interests in feeds for aquatic organisms.

Both panels had the same four assignments:

1. Answer the questions resulting from the Federal Register announcement.
2. Identify the constraints and possible solutions to providing aquafeeds in the future as fish meal and fish oil resources become scarce.
3. Identify key research and technology transfer needs to overcome barriers for reducing reliance on fish meal and fish oil resources.
4. Predict the future of aquaculture feeds based on the first three items in this list.

Case Studies

The steering committee solicited short write-ups from individuals who are already actively working to replace fish meal and fish oil. These seven case studies provide concrete examples of research leading to replacements, actual replacements being adopted by the aquaculture industry, or areas that hold great promise for replacement in the future. These case studies are listed after the summary section in this report.

Case studies included are:

1. Developing the potential of fish processing byproducts takes guts
2. From fish meal-dependent to fish meal-free: feeds research is producing the alternative diets of the future for trout
3. Plant-based feeds for black sea bass show promise
4. Shrimp farmers join with researchers to test new diets

5. Seaweed farming may be key for alternative aquaculture feeds
6. Research on diets for threatened and endangered fish species held in captivity gains ground
7. Soy products and aquaculture are a winning combination

Futurecasts

Both the researchers panel and the stakeholders panel provided the opportunity for each participant to provide their vision of the future of aquaculture feeds. At the end of each workshop, attendees were asked to spend some time thinking about and recording what they see as the state of feeds for aquaculture in the future. Specifically, participants were asked to predict the challenges and changes that aquaculture will face, and the developments that will affect both producers and consumers over the next 5 and 25 years. This information is summarized in the section on futurecasts from researchers and stakeholders panels. Because participants in each panel varied widely in background and expertise, they provided a variety of visions of the future. The unedited responses are presented following a short summary of all the futurecasts.

SUMMARY OF SCIENTIFIC EXPERTS PANEL

The purpose of the panel was to bring together scientists working in the field of fish nutrition, feedstuffs research, agriculture, biofuels, human nutrition, and byproducts processing to address a series of issues and questions regarding the future of alternative feeds for aquaculture.

Background and Organization

Research panel members were asked to work on four specific tasks as they addressed the major topic areas:

1. Help answer the questions that the public submitted based on the Federal Register notice.

2. Identify constraints and possible solutions to the question of providing aquaculture feeds in the future as fish meal and fish oil become scarce.
3. Identify key research needs for moving forward.
4. Predict the future of aquaculture feeds, based on information gathered from the first three items in this list.

This panel provided the scientific foundation for addressing the critical issues affecting the future of aquaculture feeds and how they can be addressed. This panel also addressed how to get promising research results into commercial production. In all, 21 scientists from five countries (Australia, Canada, Japan, Norway, and the United States) participated for one and a half days of meetings held at NOAA's Northwest Fisheries Science Center, Manchester Lab, in Washington State.

The workshop facilitated discussion by the whole group in five sessions focused on the following subtopics:

- Public perception, public ideas, and education needs.
- Nutrient and feedstuffs constraints and possible solutions.
- Economic and environmental impacts from alternatives.
- Human health implications of alternatives.
- Research and technology transfer needs.

Each session was approximately 3 hours long and started with a quick review of any public comments received from the Federal Register notice germane to the session at hand. This was followed by several 5-minute mini-presentations by two to five members of the panel with specific expertise in the session topic designed to stimulate and frame the discussion. Mini-presentations were followed by a moderated discussion by the whole panel. A note taker recorded key points on display paper. At the end of the discussion, panel members were invited to prioritize points recorded during the session on the display paper by placing colored adhesive dots next to the points they considered most important. Each individual panel member was given 10 dots to use to highlight discussion points and was instructed to make decisions independently. This information was used to ensure that the most important issues, solutions, and approaches were recorded.

At the end of the discussions, each panel member was assigned the following "homework" designed to provide a set of visions for the future of feeds for aquaculture:

The Future of Aquafeeds . . .

This is a take home assignment – each participant should send in within two weeks following the meeting, what they see happening in the next 5 and 25 years in the area of feeds for aquaculture. This is an exercise in science fiction so please take your best guess and use your imagination but be honest in what you really see as the future of aquafeeds. Please keep each Scenario (5 years from now and 25 years from now) to under 2 pages in length. As much as possible make them applicable to your location and species. Let us know what the diets will be composed of, what the feed efficiency and growth rates will be and what breakthroughs occur to make your scenarios possible. Where will the limiting nutrients come from and what feedstuffs will dominate the industry in your country? What species will these diets be fed to? How much aquafeed is being produced worldwide? How are these diets sustainable in the long run? You are welcome to also put in natural disasters which might affect aquafeeds.

The results of this assignment and similar responses from the stakeholder experts panel meeting are summarized in the section titled "Futurecasts from experts panels" on page 65.

Results

The discussions resulting from these sessions tended to range across subtopics, alternatives, processes, approaches, research needs, roles of governments, technology transfer, and commercial development. While this sort of open discussion was productive at generating ideas and information, it complicated organizing this report chronologically. Instead, we captured the discussion according to feedstuffs and their potential for economic, environmental, and human health performance in aquaculture diets, regardless of which discussion group was involved–shown in Table 1. Likewise, we captured the discussions on research and practical approaches to resolving issues surrounding fish meal and fish oil replacement in Table 2.

Table 1. Production, economic considerations, environmental considerations, human health implications, and potential barriers to expanded use of alternatives to fishmeal and fish oil in diets for aquaculture

Feedstuff class	Suggested by: Public, Panel, or Both	Current annual production (tons)	Economic and practical considerations		
			Cost $US/ton	Advantages	Disadvantages
Plant products	Both	~230 million metric tons	Generally between $500–$1800/ metric ton	Low cost	Incomplete nutrients
				Biotechnology can keep costs down and improve nutrients to be complete	Anti-nutrients are costly to remove
				Largest quantity of proteins and oils on the earth from plants	High in carbohydrates which are costly to remove
					Plant protein concentrates which work well for fish feeds are more expensive than fish meal

Environmental considerations		Human health/product quality considerations		Barriers to expanded use in aquafeeds
Advantages	**Disadvantages**	**Advantages**	**Disadvantages**	
Low trophic level; primary producers	Issues associated with increased agriculture	Can increase LCn-3FA by biotechnology	Naturally low in LCn-3FA	Sometime poor palatability
Can be organically produced, but will lack LCn-3FA		May contain phyto-chemicals that have positive implications for human health	May have contamination loads from farming practices (i.e. pesticides)	Can be high in anti-nutrients and carbohydrates
Coproduct provides use of waste material for another industry (e.g., starch)			May contain phyto-chemicals that have negative implications for human health	Greater processing trends to improve results with fish
Sequestors CO_2				R&D plan well documented by plant products in aquafeeds working group; should use a model for other feedstuffs where applicable

Table 1. (Continued)

Feedstuff class	Suggested by: Public, Panel, or Both	Current annual production (tons)	Economic and practical considerations		
			Cost $US/ton	Advantages	Disadvantages
Byproducts of bioenergy production	Both	For ethonol production at the end of 2008: ~26 million metric tons/year; by the end of 2015 will be between 30 and 40 million metric tons/year	Generally between $100 to $200 per metric ton	Low cost	Quality of protein sometimes low due to ethonol production process
				Use of carbohydrate fraction for bioenergy may increase availability of the protein and bring costs down	Lipid competes with biodiesel
Rendered animal products	Both	~8 to 14 million metric tons	$500 to $800/ metric ton	Use of waste material and established processes Meal is similar to fish meal in composition and has been shown in many publications to be well used by a wide variety of aquatic organisms	Public concerns over BSE have resulted in restrictions on use in feeds

Environmental considerations		Human health/product quality considerations		Barriers to
Advantages	Disadvantages	Advantages	Disadvantages	expanded use in aquafeeds
Helps make biofuels more cost-effective to increase chance for replacement of fossil fuels	Issues associated with increased agriculture	May contain phyto-chemicals that have positive implications for human health	May have contamination loads from farming practices or processing for fuels	Byproducts (DDGS) and some protein concentrates from biodiesel have poor functional qualities, high levels of indigestible material, and often poor protein quality; perhaps look at fractionation of protein before distillation or refining; variability in nutrient content, quality, and physical properties; transportation and storage challenges; sometimes poor palatability
Coproduct provides use of waste material from another industry			May contain phyto-chemicals that have negative implications for human health	
Sequesters CO_2			No LCn-3FAs	Need to work with refining process to produce higher quality byproducts for aquafeeds
Recycles animal processing wastes back to fish	Issues associated with animal production	May be a source of CLA	Some markets do not allow terrestrial animal proteins	Some regulatory issues due to BSE; Poultry byproduct meal is widely used already and regulated by costs and supply
			BSE issue is unclear	In general, these products have a great record as ingredients in diets for aquatic organisms
			No LCn-3FAs and high in other less healthy fats	

Table 1. (Continued)

Feedstuff class	Suggested by: Public, Panel, or Both	Current annual production (tons)	Economic and practical considerations		
			Cost $US/ton	Advantages	Disadvantages
Byproducts from fishery and aquaculture (fish and shellfish)	Both	~2 million metric tons (already counted as fish-meal	Same as fish meal from forage fish; $1200–$1600/ton	Easy to replace fish meal and oil with high quality meal	Expensive to capture
				High palatability	Located in small quantities from diverse sources Highly perishable until dried
Algae products (seaweeds)	Both	~1.5 million metric tons (dry); mostly for human and ruminant feed market	Depends on grade, species, and market; cost of algal protein concentrates or lipids have not yet been determined	May contain high levels of nutrients not found in terrestrial plants (LCn-3FA's, Taurine, etc.)	Low in protein and oils; would require significant processing to concentrate nutrients for fish

Environmental considerations		Human health/product quality considerations		
Advantages	Disadvantages	Advantages	Disadvantages	Barriers to expanded use in aquafeeds
Uses waste which is now discarded, often causing nutrient pollution		High in LCn-3FA and other nutrients	May have the same contaminants as conventional fish meal and oil	Difficult to process due to the temporal and spatial availability of the wastes, and their perishable nature Costs currently higher than production of fish meal and oil from industrial fishery High costs for infrastructure, drying, and transport
Low tropic level; primary producers	Requires space	May be good source of LCn-3FAs and other marine nutrients		Protein and lipids need to be concentrated (carbohydrates removed) before feeding to carnivorous fish

Table 1. (Continued)

Feedstuff class	Suggested by: Public, Panel, or Both	Current annual production (tons)	Economic and practical considerations		
			Cost $US/ton	Advantages	Disadvantages
				No need for fresh-water or land	Little grown in the US
					Competition for use as human food
Krill or wild zooplankton	Both	~120,000 metric tons wet or 35,000 metric tons dry	$2000 to $3000/ton; can be higher for human food grades	Easy to replace fish meal and oil with high quality meal	Expensive to capture
				High palatability	Located in polar regions far from where needed
					Highly perishable until dried

Environmental considerations		Human health/product quality considerations		
Advantages	**Disadvantages**	**Advantages**	**Disadvantages**	**Barriers to expanded use in aquafeeds**
Can be used to reduce and sequester CO_2 and nutrients from the ocean				Land or sea area needed for algae culture is either expensive or difficult to obtain permits in US
				Production is low and costs of production too high
Low on the food chain and selected species may support larger harvests	Suffers same issues as wild fisheries—is limited and can be overfished, etc.	May be good source of LCn-3FAs and other marine nutrients	May have the same contaminants as conventional fish meal and oil	Contains high levels of fluorine which may need removal
Represents the largest animal biomass on earth.	Supports other marine fauna and at the base of the food chain			Highly perishable needs to be processed within hours of capture
Current MSY target harvest level is higher than actual harvest level				Largely in international waters

Table 1. (Continued)

Feedstuff class	Suggested by: Public, Panel, or Both	Current annual production (tons)	Economic and practical considerations		
			Cost $US/ton	Advantages	Disadvantages
Insect Products	Both	Less than 50,000 metric tons Mostly produced for high value pet market (birds and reptiles)	Variable products on the market are higher than fish meal (up to $10,000/metric ton)	High quality protein and oil In theory, increased production would drop price below fish meal costs	High levels of non-protein nitrogen (chitin)
Single celled protein/lipids	Both	Less than 50,000 metric tons	Variable products on the market are higher than fish meal	Grown on low cost nutrients Maybe a good way to produce limiting nutrients or special molecules for aquafeeds	Typically highly capital, infrastructure, technology, and energy intensive
Marine invertebrates	Both	Less than 50,000 metric tons	Variable; products on the market are higher than fish meal	Can be grown on fish wastes and low cost feeds	Neither culture systems nor wild harvest are developed

Environmental considerations		Human health/product quality considerations		
Advantages	**Disadvantages**	**Advantages**	**Disadvantages**	**Barriers to expanded use in aquafeeds**
Can be produced from diverse waste materials			No LCn-3FAs	
Minimal direct impact to environment due to highly intensive and efficient systems		Can be a source of LCn-3FAs;		Production is low and costs of production too high
Recycles fish solids to a useful product	Wild harvest may be difficult to regulate	May be good source of LCn-3FAs and other marine nutrients	Some shellfish can contain toxins which might be passed up the food chain	Development of inexpensive culture systems

Table 1. (Continued)

Feedstuff class	Suggested by: Public, Panel, or Both	Current annual production (tons)	Economic and practical considerations		
			Cost $US/ton	Advantages	Disadvantages
		Mostly used for bait or a part of a specialized feed		Typically highly palatable to some fish; may have higher value as a palatability enhancer	Wild invertebrates may harbor pathogens and parasites to fish Cost to rear in captivity is high
Invasive species meals	Both	Unknown	Unknown and variable	May be able to generate funds for capture as well as for product	Highly variable materials (green crabs, Asian carp, zebra mussels, etc.) so considerations will differ for each type of material
				Same advantages as fishery by-products	Successful project would work it's way out of a source of product Same disadvantages as fishery byproducts

Environmental considerations		Human health/product quality considerations		Barriers to
Advantages	Disadvantages	Advantages	Disadvantages	expanded use in aquafeeds
Low on the food chain	Wild harvest may remove an important part of the near shore ecosystem			Testing as a palatability enhancer to increase utilization of other more abundant but less palatable alternatives
Provides additional incentive to remove invasive species		May be good source of LCn-3FAs and other marine nutrients *if marine species* are used	No LCn-3FAs if freshwater species	Difficult to process due to the temporal and spatial availability of the material, and it's perishability; however processes do exist to make a high quality meal from this material; high costs for infrastructure
			May have the same contaminants as conventional fishmeal and oil	Highly variable material and often difficult to harvest cost effectively

Table 1. (Continued)

Feedstuff class	Suggested by: Public, Panel, or Both	Current annual production (tons)	Economic and practical considerations		
			Cost $US/ton	Advantages	Disadvantages
Aquaculture of fish for fish-meal	Public	Unknown	Unknown	Could be a large supply not subject to limits and variations of wild populations	Likely very high cost to produce

Environmental considerations		Human health/product quality considerations		Barriers to expanded use in aquafeeds
Advantages	Disadvantages	Advantages	Disadvantages	
May reduce use of fish meal and oil from capture fishery by direct substitution	Increased use of land or ocean space	May be good source of LCn-3FAs and other marine nutrients if marine fish	No LCn-3FAs if freshwater fish	No such marine systems exist
	Would still require feeds and a source of LCn-3FAs and other limiting nutrients			Fish produced in aquaculture are suitable for higher value human market

Table 2. Research and practical approaches to resolving issues surrounding fishmeal and fish oil replacement

Issue	Approach	Options to achieve
General issues		
Lower the costs of alternative feedstuffs relative to FM and FO—the majority of alternatives are more expensive than FM and FO.	Improved sustainability must show economic benefits or a higher willingness to pay	• Research and technology improvement for alternatives • Communications and outreach with latest information • Industry should be responsible for technology development at some point—government should set limits • Need for clear administrative authority (NOAA/USDA) • Identify requirements for developing industry
Understand the environmental impacts associated with alternatives	Conduct environmental review studies	• Include review of environmental impacts of all feedstuffs in assessments • Compare to industrial fisheries and alternatives • Develop low pollution diets (N and P) by studying metabolism and absorption of feedstuffs in assessments
Maintain the human health value and eating quality of aquacultured seafood	Need LCn-3FAs for fish and higher levels for humans in final product	• Blend plant oils with fish oil or feed fish oil as finishing diet • Biotechnology-engineer plants to produce EPA and DHA • Develop low cost production of EPA and DHA from Algae and/or marine microbes • Recover more fish oil from byproducts of fish processing (wild and aquaculture) • Identify additional positive bioactive compounds in fish meal/fish oil • Beyond EPA/DHA–what are roles of SDA, ARA, and other fatty acids in humans.

Issue	Approach	Options to achieve
		• In oil replacement studies check taste, fatty acid levels and product quality • Need standard method to analyze for fatty acids in fish
	Reduce or eliminate contaminants	• Need to check for other contaminants as alternatives are used (e.g. pesticides in plants) • Monitor and keep dioxin/PCB levels low • Blend oils or filter oils to remove compounds
	Evaluate new ingredients for potential hazards	• Approach depends on alternative
	Conduct human health studies	• Link what goes in to fish to what is on plate • Studies in human populations eating fish fed alternative diets.
	Product quality	• Check flavor, texture and sensory qualities of fish fed alternatives • Should be low cost and abundant final product to increase consumption to healthy levels
Understand and manipulate the animals needs	Nutrition studies	• Improve diets for different life stages and new species • Need basic understanding of how fish use nutrients • Determine what semi-essential nutrients are in FM/FO that are needed by aquaculture organisms • Determine semi-essential nutrient levels that optimize performance (e.g. taurine)

Table 2. (Continued)

Issue	Approach	Options to achieve
	Genetic studies	• Species x nutrient interactions • Understand and use genetic diversity in cultured organisms
	Fish health studies	• Testing alternatives for impacts on health and intestinal morphology
	Fish physiology studies	• Improve our understanding of nutritional physiology in fish • Increase nutrient retention • Understand food allergies in aquaculture organisms • Determine how fish metabolize fatty acids–fates of ALA, EPA, DHA, SDA, etc
	Fish physiology studies (continued)	• Explore use of n-3 FA's as biomarkers. • Develop finishing diets and models of fat metabolism • Do longer term studies in fish
	General	• Understanding functional properties of ingredients
Pellet quality needs to be maintained		• Understand impacts on texture and palatability of pellets • Increase effort to develop those alternatives that may increase under climate change scenarios
Understand the impacts of climate change on feed-stuff quantity and quality	Approach depends on alternative	• Need for long term funded research • Scientific collaboration

Issue	Approach	Options to achieve
Research and development needs to be increased and improved	Improve support, usefulness, and efficiency of research	• Industry involved
	Develop a database with info in one place	• With cost, composition and formulation information
Improve communication with human nutrition community	General	• Publish in Journal of Nutrition and trade publications • Attend and present at human nutrition conferences
	Develop a Risk/benefits model for aquaculture products	• Develop and populate a risk/benefit model for fish consumption with data from farmed fish
Education and outreach to consumer and public needs to be increased and improved	Conduct human health studies	• Conduct a "lightning rod" study to demonstrate benefits
	Demonstrate benefits of farmed and wild fish consumption	• Address public perception of value of pills vs. food • Highlight positive role of fish in diet of children. • Highlight seafood's role in fighting obesity
	Conduct economic studies	• Conduct break even/willingness to pay studies. • Ensure consistent and stable source of supply of feedstuffs • Demonstrate sound economic models • Understand timing of when products are available
	Increase visibility of aquaculture	• Increase public relations efforts • Attend meetings outside of aquaculture area. • Get aquaculture and fisheries working together. • Reverse perception that aquaculture products are unhealthy • Put aquaculture scientists on USDA grant panels. • Demonstrate nutritional benefits of aquaculture products

Table 2. (Continued)

Issue	Approach	Options to achieve
		• Partner with food scientists and human nutritionists
	Flavor of product for humans maintained	• Partner with economists and business experts
	Cost to consumer needs to be low	• Partner with social scientists
	Market issues with alternatives need to be understood and addressed	• Investigate technologies to improve feedstuffs in general
Improve processing options to improve quality and reduce costs		• Develop low cost/low energy methods to dry and stabilize meals • Increased use of air classification
Issues specific to different alternatives		
Increase the use of plant based feedstuffs	Plant protein concentrates price needs to drop	• Improve technology and develop industry. • Improve crop and coproduct consistency and quality
	Follow the PPA strategic plan (Options taken directly from goals of the "Plant Products in Aquafeed Working Group Strategic Research Plan")	• Establish standardized research approaches and protocols for systematic evaluation of plant feedstuffs across carnivorous fish species • Enhance fish germplasm and discover genes • Enhance the inherent composition of crops to provide a beneficial balance of bioactive compounds in order to optimize their use in aquafeeds for carnivorous fish

Issue	Approach	Options to achieve
		• Increase understanding of interactions between gastrointestinal microflora and plant tolerance in fish • Improve and optimize ingredient processing, feed manufacturing technology and feed formulations to increase inclusion of plant-derived ingredients in the diets of carnivorous fish • Optimize the storage, nutritional and sensory quality of aquaculture species for human consumption • Develop an international communications network fo research on optimizing plant products in aquafeed
	Understand plant composition as it relates to fish requirements	• Determine what is missing in plants • Conduct taurine supplementation (and other semi-essential nutrients) and metabolism studies to improve plant based feeds
	Enhance plant fatty acids	• Use biotechnology
	Enhance plant proteins and products	• Improve processing • Use conventional plant breeding • Use biotechnology • Determine all the anti-nutrients in plants • Cure induced enteritis in salmonids
	Develop high value molecules from plants	• Engineer plants to express carotinoids, etc.

Table 2. (Continued)

Issue	Approach	Options to achieve
Increase the use of bioenergy coproducts	Increase economic and nutritional value of coproducts	• Examine processes and develop alternatives that produce higher quality coproducts • Protein research • Fiber research • Improve digestibility of DDGS • Improve crop and coproduct consistency and quality
Increase the use of marine algae	Develop high value compounds first.	• Investigate as a taurine source • Improve economics of DHA/EPA production by algae • Develop and improve fractionation techniques to increase protein/reduce fiber contents
	Develop easy methods to concentrate proteins and lipids	• Investigate enzymatic processing to concentrate protoplasts/reduce fiber contents • Investigate mechanical processing to concentrate protoplasts/reduce fiber contents
Increase the use of sustainably harvested krill and zooplankton		• Support ecosystem-based management of fisheries for feedstuff production • What is harvested for feed should be low on tropic level for increased biomass • Determine benefits and risks associated with a fishery for krill
Increase the use of cultured marine invertebrates		• Explore economic and nutritional value of cultured zooplankton meal (e.g. rotifers) • Explore economic and nutritional value of cultured invertebrate meal (e.g. Polychetes)

GLOSSARY

ALA	Alpha-linolenic acid (18:3n-3)
ANF	Anti-nutritional factors
ARA	Arachidonic acid (20:4n-6)
BSE	Bovine spongiform encephalopathy
CLA	Conjugated linoleic acid (18:2n-
DDGS	Distillers ain and solubles (byproduct dried gr
DHA	Docosahexaenoic acid (22:6n-3)
EPA	Eicosapentaenoic acid (20:5n-3)
FM	Fish meal
FO	Fish oil
HUFA	Highly unsaturated fatty acids
LCn-3FA	Long chain omega-3 fatty acids A and DHA)
N	Nitrogen
OMP	Oregon moist pellet
P	Phosphorus
PBM	Poultry byproduct meal
PCB	Polychlorinated biphenyls
POP	Persistent organic pollutants
PPA	Plant Products Aquafeeds Working Group in
PUFA	Polyunsaturated fatty acids
SDA	Stearidonic acid (18:4n-3)

SUMMARY OF SCIENTIFIC EXPERTS PANEL

Results

Feedstuffs

It is noteworthy that the cost per unit of protein for fish meal and per unit long chain omega-3 fatty acid (LCn-3FA) for fish oil is low relative to alternatives (Table 1). It is also clear that when a substitute provides a cost advantage to the overall diet without affecting performance it is quickly adopted by feed manufacturers. Indeed, a modern commercial salmon diet that may contain 30 percent fish meal is already 70 percent something else. Research, development and technology transfer that can help reduce the costs of abundant alternatives relative to fish meal and fish oil will likely result in quick incorporation into commercial diets.

One metric that may be a good predictor of future substitution potential is the amount of the protein or oil source available on the world market and its price. Typically, world market production and price are inversely correlated to a certain extent. The major protein sources are shown graphically in Figure 3 and the major oil sources in Figure 4. Ranges for this information are also given in Table 1. Clearly, much of the world's supply of protein and oil comes from plants. Plants already supply the majority of the protein and oil found in feeds for aquaculture, and this trend will likely continue. Plant products specifically mentioned by the panel include soy, canola (rape), corn, wheat, cottonseed, lupin, sunflower, flax linseed, and peas. Protein and oil concentrates, and gluten meals made from these crops would be the most useful for aquaculture, while the carbohydrate fraction is not useful.

Since biofuel production for alcohol, and bioplastics production for degradable plastics, utilizes the carbohydrate fraction of plants to make ethanol and other molecules, there was a lot of interest in the potential synergies between these industries and feedstuffs for aquaculture. Biodiesel production uses the oil fraction of plants, but this could create high protein concentrates as byproducts that could suit aquaculture diets. Since the bioenergy and bioplastics industries are likely to be orders of magnitude larger than aquafeeds in the foreseeable future, the panel recommended working with these industries to ensure high-quality protein and oil by-products that would be suitable for aquaculture feeds.

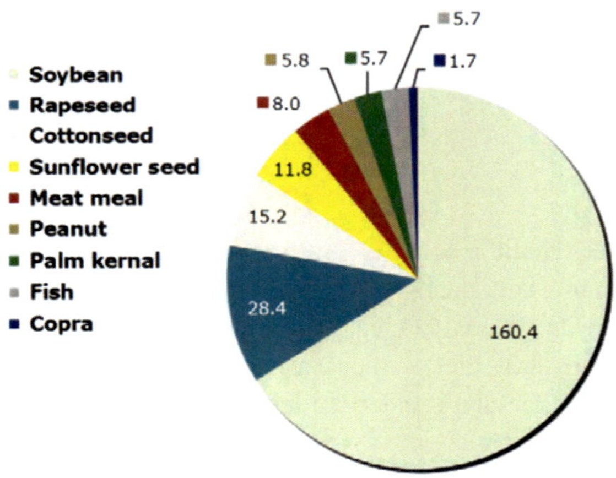

Figure 3. World production of protein meals in 2007 (millions of metric tons).

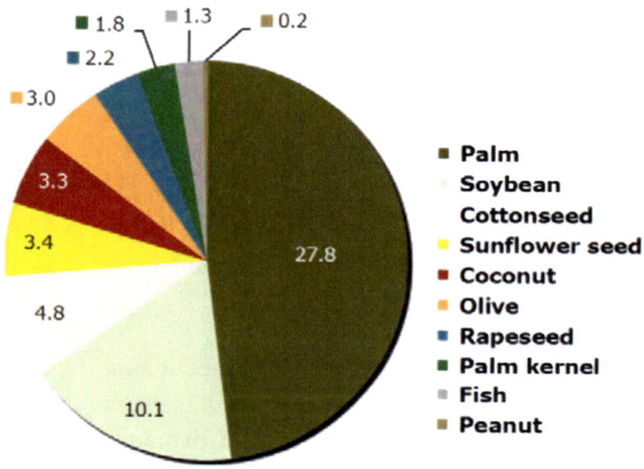

Figure 4. World oil production in 2007 (millions of metric tons).

Prior to this initiative, an ad hoc group of researchers formed the Plant Products in Aquafeeds Working Group (PPA). This group has published a review paper (Gatlin et al. 2007) a strategic plan (Barabout.php) focused on increasing the use of plant products for aquaculture feeds. Much of this is directly applicable to the NOAA/USDA Alternative Feeds Initiative. The scientific panel reviewed and endorsed the approach and planning by the PPA in their strategic plan, and those suggestions have been incorporated in Table 2. There are also extensive comments from the PPA available on line in the public comments page (http://aquaculture.noaa.gov/news/comment.html.).

Other feedstuffs considered include byproducts of the fermentation industry (especially distillers dried grains and solubles—DDGS), byproducts from the capture fishery and aquaculture, algae, krill, zooplankton, insects, single celled protein (SCP—mostly yeasts and bacteria), marine invertebrates (mollusks and polycheates), rendered animal products (meat and poultry byproduct meals), and invasive species meals. An additional suggestion made by the public was to rear species by aquaculture as a source of fish meal for aquaculture. The economic, environmental, and human health implications of these alternatives are summarized in Table 1 along with some of the challenges they present for fish feeds.

Nutrients

To a certain extent it was difficult to separate the discussion of nutrients from the discussion of the feedstuffs that contain them. However the need by fish for specific nutrients, rather than for any one feedstuff per se, provided some clear focus. In terms of identifying key nutrients, two classes of nutrients were identified as key to the future of feeds for marine aquaculture—long chain omega-3 fatty acids (LCn-3FAs) and compounds that are known and unknown in fish meal and fish oil that act as semi-essential nutrients (for example taurine, perhaps cholesterol, hydroxyprolene, phospholipids, and others). In addition, the levels of known essential nutrients that might need to be adjusted due to protein and oil substitutions (such as vitamins and minerals) that change either the requirement or bioavailability of these nutrients, needs to be investigated.

Two LCn-3FAs occurring in seafood—eicosapentanoic acid (EPA) and docosahexanoic acid (DHA)—are well known for their human health benefits. The panel repeatedly emphasized the importance of maintaining levels of these fatty acids in the products of aquaculture, primarily for human consumption and secondarily for fish, since fish tend to require a lower level than that which would be optimal for human health. Currently, the majority of LCn-3FAs in aquaculture diets are obtained from fish oil. However, forage fish do not make these fatty acids themselves but rather concentrate them through the food chain. The primary producers of these fatty acids in nature are marine algae and microbes. Two basic approaches to reducing reliance on fish oil for EPA and DHA were described:

1. Alter the feeding approach by growing fish on low LCn-3FA diets and use a “finishing” diet to boost these fatty acids at the end of the production cycle, thereby making the use of fish oil and other oils containing EPA and DHA more efficient.
2. Produce DHA and EPA from primary producers (algae and/or other marine microbes) or genetically modified plants, fungi, or microbes, and then replace some or all of the fish oil with blends of the alternative oils in the diet.

The panel agreed that both approaches should be investigated. The functional role of additional long chain fatty acids in fish oils are unknown but should also be investigated.

For the second class of nutrients identified by the panel—the semi-essential nutrients that are found in fish meals and fish oils but perhaps not in all alternatives—the approach suggested is to first identify all of them and then find ways to replace them. For this discussion, a semi-essential nutrient is one that is not needed for growth and survival per se, but is required for maximum growth rate, disease resistance, or other performance traits desirable in aquaculture. Repeatedly used as an example was taurine, a sulfur-containing amino acid not found in plants but abundant in animal tissue. The addition of taurine to plant proteins to make a higher performing diet is not difficult and abundant sources of taurine exist from animal by-products, some algae, and synthetic sources. However, the story of taurine illustrates the possibility that other semi-essential nutrients may not have been discovered, but may be lacking in alternative feed ingredients. The panel urged further research to discover additional semi-essential compounds so they can be incorporated into aquaculture diets.

The opposite issue pertains to anti-nutrients that may be abundant in alternative feed ingredients. Anti-nutrients are compounds that negatively impact the health or nutrition of the consuming animal. There are many anti-nutrients in plants, which have evolved to contain these compounds as a defense against grazers. An example is trypsin inhibitor found in soybeans that allows them to pass through an animal's gut less digested and able to germinate. While plants are the classic example of meals containing anti-nutrients they are also found in other types of meals. Many of these anti-nutrients are broken down by heat or removed in the processes used to concentrate protein. The panel recommended further work on identifying and developing efficient processes to remove or destroy anti-nutrients in feed ingredients.

Looking beyond the practical and economic considerations of various feedstuffs, the panel also considered the environmental footprint of potential substitutes. Environmental impacts associated with feedstuffs were discussed in general terms for classes of feedstuffs from two points of view; 1) the environmental impacts associated with procuring it and 2) the environmental impacts of using it.

All of the proposed potential protein and oil sources could be sustainably developed to some extent. Wild harvest of algae, krill, zooplankton, and worms, as well as the processing of wild fish trimmings, still relied on good management of a wild resource, but generally represented the harvest of organisms lower on the food chain than traditional fish meal resources, or were from organisms already harvested for human consumption. The culture of algae and plants for feedstuffs presents an opportunity to increase supply beyond the limits of wild production by the application of aquaculture and agriculture; the former requires not much more than ocean surface area and management, whereas the latter requires all the inputs common to agriculture. Both primary producers have the added environmental benefit of capturing carbon. Culture of insects and marine worms provides an opportunity to produce protein and oils from materials fed to these organisms that would otherwise be discarded. The use of materials from bioenergy and bioplastics provides the additional benefit of helping to improve the economics of those industries that can reduce reliance on fossil fuels. The use of rendered animal products and fish trimmings from aquaculture and wild harvest would help recycle high-quality protein and oil and keep this material out of landfills or the environment. Invasive species meals would also provide incentive to remove an invasive species, with presumed ecological benefits to the invaded ecosystem; however, the likely goal of such a program would be to reduce the supply of this material over time to very low levels, making its long-term economic future uncertain. Single-celled proteins and oils can be grown on very simple substrates and may be practical to supply key limiting nutrients or bioactive compounds that would enable the use of other alternative feedstuffs in aquaculture feeds.

The primary considerations for the environmental impacts of feeding alternative feedstuffs once the diet is complete are the changes in the amount of nitrogen (N), phosphorous (P), and/or solids produced by fish fed alternative feeds. Since this concern is different depending on the nature of the receiving water and the types of feedstuffs used, it needs to be considered within context. Dietary nutrient efficiency should be investigated as promising dietary changes are discovered. The panel urged review of the environmental impacts of alternative feedstuffs relative to industrial fisheries and studies to understand impact to N and P metabolism and solids production.

The final area of consideration for substitutions was their impact on the quality of the product. Product quality for fish produced for human consumption has two components: 1) the impact on the health of the consumer, and 2) the impact on the taste, texture, and look of the product as food. Both areas were considered important, but the panel focused most of its attention on the first component. Fish are a healthy choice for human consumption because they contain essential and high-quality protein, oil, minerals, and vitamins. While all of these areas require monitoring to some extent when substitutions are made, it is primarily the oil and oil-soluble vitamins that can be altered due to diet. Therefore, the majority of the panel discussion focused on the oil fraction. Within the oil fraction, the majority of interest was on the n-3 fatty acids. On the other side there was discussion on the need to reduce potential contaminants such as methylmercury and PCBs (and other persistent organic pollutants—POPs). Alternatives generally have the potential to reduce the heart-healthy n-3 fatty acids unless specific attention is paid to including EPA and DHA in the diet. On the other hand, many of the alternative feedstuffs are lower in contaminants than fish oils. One consideration for plant meals is to examine the impacts of residual pesticides used in the farming of the replacement plant meal. Specific recommendations are presented in Table 2.

Alternatives to fish meal and fish oils have already come a long way. For example, in the 1960s diets for salmonids contained 60 to 80 percent fish products (based on Oregon moist pellet [OMP] and Abernathy open formula diets), while today's modern commercial salmonid diets contain only 30 to 40 percent or less of fish products, and this percentage continues to drop. It is clear that the use of alternatives has become cost-effective and this trend will continue. Research to reduce the barriers and cost to use alternatives needs to expand and keep in mind the environmental and product quality considerations that substitution and widespread use of those alternatives bring. Researchers in fish nutrition and alternative feedstuffs development should increase communication with scientists in the human health, agriculture, food sciences, bioenergy, aquaculture, animal physiology, and environmental sciences to help evaluate alternatives. This collaboration can help meet the goal of developing sustainable feeds for economically profitable, environmentally sound, and high health product aquaculture.

SUMMARY OF STAKEHOLDER EXPERTS PANEL

National Stakeholders Meeting

In April 2008, as part of the Alternative Feeds for Aquaculture Initiative, the agencies held a national stakeholder meeting in Silver Spring, Maryland. The purpose of the meeting was to provide a forum for open communication among stakeholders including scientists, representatives from government and non-governmental organizations, academia, private industry, and others regarding trends, opportunities, challenges and key issues related to the development of alternatives to fish meal and oil in aquaculture diets. The meeting attracted over 60 participants. Four major topics were addressed, including human health and product quality, environmental implications, alternative feedstuff options, and future directions for feeds manufacturing. This section is intended to provide an overview of the stakeholder meeting, including the expert presentations and stakeholder discussions.

Hosted by NOAA's Aquaculture Program, the meeting was brought to order by Dr. Michael Rubino, the agency's Aquaculture Program Manager. Dr. Rubino set the stage with background information on the pressing need for the rapid development, testing, and commercialization of alternative feed ingredients. He also presented the challenge to the group from the standpoint that the environmental and financial sustainability of aquaculture depends on developing alternatives to fish meal and fish oil as the primary feed ingredient. Furthermore, he stressed that the nutritional characteristics of fish meal and fish oil-based feeds, including essential fatty acids, are integral to what make fish and shrimp a healthy food choice from a human health perspective. He also posed several critical questions to the group, as well as to the broader aquaculture and seafood communities and policymakers, including:

- Which alternatives are most promising?
- Which deserve our attention?
- Which deserve our federal research dollars?

The meeting was moderated by Dr. Paul Sandifer, then the Senior Scientist for Coastal Ecology for NOAA's Ocean Service. Dr. Sandifer is now the acting-Senior Science Advisor to NOAA Administrator Dr. Jane Lubchenco. He set the tone for the meeting with the following observations:

- It is clear the increasing human population has an increasing demand for seafood, and even in the most optimistic of scenarios, we have a fully developed wild fishery situation, stable or declining. There simply is not much more we can take from the oceans.
- We need aquaculture, but we need aquaculture to be a sustainable green industry—to have as little negative and as great a positive impact as possible.
- The concern is that increasing demand could put unsustainable pressure on wild capture fisheries.
- There is also the question of the economic sustainability of aquaculture -increasing costs and increasing demand will continue to result in higher costs and more competition for fish meal and fish oil.
- In aquaculture we are dealing with controlled feeding, not the ability of animals to graze in the natural environment. We need to take advantage of this. Rather than using more and more wild capture fishery products, we need to figure out how we can better supply the essential amino acids and essential fatty acids from other ingredients.

- There is considerable difference of opinion about the best uses for the natural feed ingredients. The bottom line is that unless alternative feedstuffs are found, demand for aquaculture will surely outpace affordable supply. So whether you look at this from the environmental or ecological perspective or you look at it strictly from the business perspective—better yet, look at it from both perspectives—the development of alternatives to fish oil and fish meal in aquaculture feeds is essential.

Providing the USDA research perspective on alternative feeds, Dr. Caird Rexroad, Associate Administrator for the USDA's Agricultural Research Service (ARS), presented an overview of the agency's commitment to aquatic animal nutrition including alternative feeds. He discussed challenges, issues of competition, and options to advance the field of aquaculture nutrition.

He noted that the current competition between food and fuels is an issue that is just now being felt by Americans, and this issue will continue to place challenges on future uses of feedstuffs as well as the economics of agriculture and aquaculture. He also noted that a key scientific tool to address many of the feeds issues is genomics. . . . "(genomics) will be a big part of anything that we use as we try to understand nutrition and the response of these species in breeding to alternative feed sources."

Dr. Ralph Otto, Associate Administrator of the Cooperative State Research, Education and Extension Service (now the National Institute of Food and Agriculture, NIFA) opened his remarks about the variety of strategies required to address the very real challenges and changes facing agriculture and aquaculture on a recurring basis. This included the critical need both domestically and globally to identify and incorporate new and practical alternatives for fish meal and oil in feeds for aquatic animals. He stressed the added challenge of maintaining and even enhancing the human health benefits of consuming farmed aquatic foods as well as protecting sensitive natural resources associated with aquatic production sites. He also noted that successful strategies are developed at these types of meetings that join people of diverse interests, knowledge, and even sometimes competing interests, but keenly focused on pathways for solutions rather than problems and obstacles.

He stated that NIFA is a grant funding agency working across the broad spectrum of agriculture and in partnership and support of the land grant university (LGU) system with research, education and extension programs, some of which include aquaculture. NIFA also shares resources and people with ARS through numerous co-located laboratories at LGU and also with NOAA Sea Grant extension programs.

He explained that NIFA recently expanded integrated programs that include research, extension and/or education to mobilize expertise across these functional programs on distinct problem areas taking full advantage of multidisciplinary and collaborative approaches for solutions that equate well with the challenges that will be addressed at this meeting. He stressed a real challenge is how to most effectively focus basic research on the most critical problems and efficiently translate increased understanding of complex systems and processes to applied research and practical applications cost-effectively. Science moves our knowledge system forward and can create new visions and possibilities for today and in our future. Dr. Otto asked the participants to imagine a world where foods from aquaculture not only provide fundamental nourishment but also improve our day-to-day health and contribute to clean water, clean air, and a sustainable environment. He noted that the US Department of Agriculture was created by President Abraham Lincoln and in his last public address he reminded us that important principles may and must be flexible. Dr. Otto challenged the participants to be flexible and move forward in a way that anticipates not only what is but what can be as new ideas emerge to stimulate critically needed progress and solutions on alternatives to fish meal and oil in aquatic animal feeds.

Following these opening remarks, four overarching issues were introduced by recognized experts including Dr. Charles Santerre of Purdue University (human health and product quality), Dr. Jane Lubchenco then of Oregon State University and now the current NOAA Administrator (environmental implications), Dr. Diane Bellis of Ag Source, Inc. (alternative feedstuff options), and Mr. Richard Nelson of Silver Cup Feeds (future directions for feeds manufacturing). Following the introductory presentations, breakout groups for the four issues provided focused discussion on specific challenges, status, needs and research priorities, and mechanisms to facilitate progress.

Human Health and Product Quality Impacts

Dr. Charles Santerre provided an overview of the health benefits of seafood and a human health perspective of the role of seafood in diet. Highlights from Dr. Santerre's presentation included the following points:

- The old saying "you are what you eat" applies to fish as well. The nutritional value of fish is based on their diet. What is in the diet of fish will be ingested by humans.
- The nutrients in fish include the healthy omega-3 fatty acids (EPA and DHA), selenium, calcium, iodine, zinc, vitamin D, arginine, conjugated linoleic acid, and polyphenols among others;
- Fish ingest these key nutrients as part of their diet and then pass them on to humans. We need to pay attention to the impact of changes to fish feed because those changes may impact human health.
- Three and one-half ounces of salmon contains 90 percent of the recommended daily allowance (RDA) of vitamin D for an adult and fish can be an excellent vehicle for getting vitamin D into the human diet.
- DHA is one of the essential omega-3 fatty acids and is important early in life for healthy brain and eye development. Later in life, these nutrients are also important.
- Drs. Dariush Mozaffarian and Eric Rimm of the Harvard Medical School published the results of a landmark study in 2006 that determined the impact of fish consumption on human heart health looking at mortalities from sudden cardiac death. The study found that the long chain omega-3 fatty acids (EPA and DHA) contained in fish protect the heart during a heart attack. Their work found that one to two servings of fish per week; especially species higher in EPA and DHA, reduces the risk of coronary death by 36 percent and reduces total mortality by 17 percent.
- Based on the American Heart Association's estimate of more than 300,000 deaths every year due to sudden cardiac death, 120,000 lives could be saved every year if people would consume more fish or fish oils that contain the omega-3 fatty acids, EPA and DHA.
- Omega-3 fatty acids are important for our brain health as we age. One study, published by Dr. Martha Clare Morris at the Rush Institute, showed that individuals consuming one fish meal per week had a 10 percent slower cognitive decline. Further, those consuming two fish meals per week had a 13 percent slower cognitive decline.
- Americans are currently eating about 16 pounds of seafood per person per year, which is about half the amount needed to realize the full benefits.

- Nutritionists keep encouraging consumers to eat 8 ounces of seafood per person per week—or two seafood meals per week—to gain the health benefits.

Dr. Santerre noted that the potential and the challenge for aquaculture is supplying healthy seafood in an economically and environmentally sustainable manner. Increasing the array of suitable feed ingredients will improve the stability of supply and the sustainability of aquaculture.

Dr. Santerre's recommendations to achieve greater supply and sustainability of healthy seafood included:

- Improved nutrient content information for all seafood.
- Use of biotech crops designed to contain DHA and EPA.
- Need for fish nutritionists, human nutritionists, food scientists, and others to work collaboratively to explore farmed seafood as a "functional" food.
- Increased amount of fish in the marketplace, especially those species that provide the important nutrients.

Environmental Implications

Dr. Jane Lubchenco provided a new context for considering what 'sustainable aquaculture' means. This context begins with an overview of the concept of ecological services and the alterations to the provision of those services that result from various human activities. In her presentation, she described the findings of the Millennium Ecosystem Assessment (MEA), which evaluated the benefits provided to people by the functioning of ecological systems, and how society is doing in maintaining these critical services. These benefits arise from the interactions among species and between species and the environment. Examples include food, climate regulation, purification of air and water, and protection of coasts against storms, among others. The MEA concluded that 60 percent of the ecosystem services for which enough information exists are in decline. A key example of an ecosystem service in decline is capture or wild-caught fisheries. Agriculture and aquaculture are two of the services on the increase, but, according to the MEA, their increase is coming at the expense of other services.

The modification or conversion of an ecosystem results in a tradeoff of ecological services—e.g., when mangrove ecosystems are developed for shrimp production or another form of agriculture. An intact mangrove ecosystem provides nursery habitat critically important to many species, seafood, fuel, and timber. In addition, mangroves trap sediment and thus protect downstream coral reefs from sedimentation. And, as demonstrated by, the Indian Ocean tsunami, mangroves protect shorelines against erosion, absorbing the energy of storms. As mangroves are converted to homes, shrimp ponds, or agricultural crops, certain services are gained and others lost. This knowledge suggests that shrimp aquaculture needs to be viewed in the larger context of not only what is gained (seafood, jobs), but what other ecosystem services are lost with the construction and operation of shrimp ponds.

According to Dr. Lubchenco, "It is important to understand that it's not just a particular activity that needs to be sustained through time, but that the inputs to that system and the outputs from that system need to be sustainable in the larger context. It's not just about growing more of something; it's about doing it in a way that does not negatively impact the provision of other ecosystem services."

Dr. Lubchenco suggested several possible solutions to helping to make aquaculture more sustainable, including:

- Consider all the implications of any feed source—for example, shifting to by-catch of non-target species might be problematic for the same reasons that fishing on small pelagics more and more is problematic, in terms of the disruption it would likely cause to many food webs.
- Terrestrial plants, bacteria, microalgae, protists, and yeasts are worth exploring.
- Better use of seafood products that are already being processed.
- Perhaps marine invertebrates, i.e., marine polychaetes (worms).

Alternative Feedstuffs Options

Dr. Diane Bellis discussed the necessity and opportunities for development of alternative aquaculture feeds, and provided an overview of the Plant Products and Aquafeed Working Group. The group is a coalition of researchers with a goal of coordinating research, exchanging data, and establishing standards for research on plant-based feeds. In her presentation, she stated that the demand for alternatives to fish oil and fish meal will continue to be strong, even when the demand for food and energy creates competition globally. She also stated that soybeans have been and will continue to be a part of future efforts in increasing alternative feedstuffs in aquaculture feed. A number of characteristics and factors of soy make it an important feedstuff:

- Soy has the best balanced profile among plant proteins.
- Soy accounts for 70 percent of the world's protein meal consumption.
- In the United States, 98 percent of the soybean meal is used in animal feed, primarily for poultry and swine.
- Aquaculture is the fastest growing segment of the market for U.S. soybeans.
- The United States exports about $9 billion worth of soybeans annually, nearly equal to the amount of the U.S. seafood trade deficit.

Challenges pertaining to soybeans and their use as an aquaculture feed ingredient also exist. Ninety percent of soybean oil is currently used for human consumption, leaving 10 percent for other uses including aquafeeds. It boils down to an economic and human nutrition issue as to use of oils. Development of alternative feedstuffs and sustainability will be a function of well-funded, strategic, highly credible research. Development must be neutral or beneficial to the ocean ecology, economically viable, and technically feasible. The road to this development will be rocky. We know relatively little about 25,000 species of fish. There are about 3,000 marine species that we eat, and 220 fish species of economic interest. The variation of nutritional needs is extremely wide and only a few fish species have been bred for production. Furthermore, those species have yet to be bred for improved efficiency at digesting any protein, let alone plant or other alternative proteins.

Progress has been made, but results are most often limited and confusing and come at a high price. All the research tools and breakthroughs have yet to be used effectively and the investment must be expanded to effect tangible progress. We need to learn to feed new fish species efficiently and sustainably–thus the Plant Products and Aquafeed Working Group and their efforts.

The working group's first effort was a review paper that formed the basis for the plant products and aquafeeds strategic research plan, which outlines seven goals with performance measures:

1. Improve the quality of reporting on feeding trials, the different techniques used, and formation of standard approaches and protocols;

2. Use genomics to match feed with animals;
3. Improve plant quality;
4. Explore gut microflora and probiotic aspects;
5. Processing to optimize the efficiency of feeding these plant-based diets;
6. Expand marketability of the product–characteristics for human consumption; and
7. Keep the organization moving and making the data available to people who can use them.

Soybean farmers are committed to these efforts and have funded three projects: development of a control line of fish with known genetics; a large feeding trial feeding low phytate soybeans to fish; and a synthesis of the literature to identify the gaps. Dr. Bellis recommended that the NOAA-USDA Alternative Feeds Initiative include the following:

- Develop a road map for identifying the research needed beyond plant-based diets, for including other alternatives, and build on the process that the PPA has developed.
- The Joint Subcommittee on Aquaculture can play an increased role in this effort, and it is critical that all federal agencies having a stake in aquafeeds be actively involved.

Feed Manufacturing Challenges

Richard Nelson, of Nelson and Sons, Inc. highlighted the fact that fish farmers focus on four basic and interrelated elements:

1. Create a healthy animal in its environment;
2. Grow that animal to a market size and sell it;
3. Deal with the metabolic waste that occurs on the farm that can have an environmental, regulatory, and water quality impact; and
4. Be able to make it all work economically—the farmer doesn't want to go into the business of losing money or breaking even.

Mr. Nelson noted that for each emerging species or every known species, a tremendous amount of work goes into the research to understand the nutrient requirements of that fish. Algae-based DHA oils are of much interest and ". . .

guys like me are on the edge of our seats waiting for this to come out onto the marketplace and be produced at such a rate that the economics of it will make sense." He also noted that the feeds industry is moving toward increasing the nutrient density of the feeds to lower the food conversion ratios (FCRs). "The better we do that, the better growth we get, and the less metabolic waste we introduce into the environment," he said.

"We have been practicing the art of alternative and replacement proteins for 30 to 40 years. What has driven the industry to make changes in terms of the use of fish oil and fish meal hasn't been driven by environmental pressure, and it hasn't been driven by regulations; it's been driven by *economics*."

Breakout Groups

Following the expert presentations, the participants self-selected into four breakout groups for discussions. The four topics were:

- Human health and product quality, led by Dr. Michael Rust;
- Environmental implications, led by Dr. Jeff Silverstein;
- Alternative feed stuffs, led by Dr. Rick Barrows; and
- Manufacturing challenges and future directions, led by Charlotte Kirk-Baer.

Human Health and Product Quality

The charge to this breakout group was to address the following questions:

1. What aspect of seafood's effect on human health is most important to you or your constituents or the consumer? How important is seafood as an alternative to red meat for cardiovascular benefits, for neurological development in children or other health benefits. How might alternative feeds affect this concern?
2. What aspect of seafood's product quality is most important to you? What is the relative appeal on nutritional health benefits versus the perceived risks form chemical contamination and toxins? How might alternative feeds affect this concern?
3. What is the best way to develop and communicate information on health benefits and product quality of seafood reared on alternative feeds to enhance consumer confidence? Who should be developing

and communicating this information to the consumer – industry, NGO's, government, academia, or collaborations of these?

4. What makes seafood unique in delivery of healthy compounds (high protein, low saturated fats, high omega-3 fatty acids, vitamins)? Will alternative feeds reduce the appeal of seafood?

The group discussed and outlined key questions or needs to address regarding human health and product quality as follows:

Key Health Considerations

- Change in fatty acid profile, n6:n3
- Se, Zn, micro and macro minerals (except iron)
- Supplements vs. fish
    - Benefit analysis
    - Affordability
    - Digestion problems
    - Other nutrients
        i. Taurine
        ii. Astaxanthin
        iii. Vitamin D

"Designer Food" to Target Specific Health Issues–Priorities

- Collaboration among fish nutrition, food safety and human nutrition
- Establish database of feed ingredients
- Development of functional foods
- Maintaining/increasing supply of fish
- Development/use of models to explore impacts of diet shifts in
    - Human consumption of fish on human health
    - Fish diets on human health
- Develop reference list of nutrient composition of feedstuffs

Product Quality

- Taste, palatability, texture
- Cultural identity–indigenous species
- Shelf life

Environmental Implications

This breakout session was challenged to address the following questions:

1. What are the key metrics for alternative protein and oils (costs, availability, and carbon footprint)? How can we incorporate value of the environmental impact or lack of impact?
2. Of the possible alternatives to pelagic fishery derived proteins and oils, are there some that are more appealing form an environmental perspective? Who should take the lead to develop these alternatives?
3. Can we identify benefits to the environment that are direct and or indirect of reducing the use of pelagic fishery derived proteins and oils?

The group focused on the metric relating the decision process for evaluating environmental impacts of feeds and alternative sources of protein and oils:

Alternative Sources of Protein and Oils Key Metrics

- Product quality–omega-3 fatty acids
- Feed efficiency
- Suitability–nutrients
- Species–specific needs

Specific Environmental Questions/Concerns Were Listed and Discussed Including

- Processing wastes
- Economic feasibility
- Long-term impacts
- Ecosystem bioaccumulation
- Local availability–compare domestic vs. foreign
- Major vs. supplemental ingredients
- Complete replacement?
- Consequences of genetically selected organisms/interactions
- Ecosystem impacts of nutrient extraction
- Perception of environmental impact
- Carbon footprint analysis
- Conversion ratios of poor alternatives e.g. insects minimize inputs
- Byproducts–reuse, recycle, etc.
- Toxins, contaminants, excess nutrient release (e.g. phosphorus in Idaho)
- Standards/protocols for feed trials (e.g. waste)
- Consumer Information

- Framework for evaluating environmental impacts
- Life cycle analyses
- Genetic diversity
- Green house gas and equivalents
- Regulatory solutions
- Energy exchange

Alternative Feed Ingredients and Approaches Could Include

- Determine nutrient requirements
- Use palatability enhancers
- Selective breeding for enhanced utilization of alt. ingredients
- Integrated aquaculture/polyculture
- Aquatic microalgae
- More consistent quality products
- Vegetable oil, omega-3 sources for taste and health benefits
- Marine worms-zooplankton–protein and oil
- Bacterial meal from water stream of food processing
- Converting fuel quality oils to feed quality
- Insects and insect meals
- Essential fatty acid oils (GMO)
- Micro algal and crops
- Black soldier fly meals
- Public perception and barriers to acceptability of alternative ingredients
- Borage oil
- Yeast base and proteins
- Scavenging long chain fatty acids from algal biodiesel
- Poultry byproducts
- Waste streams of poultry processing
- Organic sources "natural"
- Compositional data for alt. ingredients
- Plant proteins and plant proteins concentrate
- High-DHA algal meal protein & oil
- True value of EPA and DHA to fish
- Byproduct breweries, wineries, farms, and coproducts of biodiesel
- Byproduct from fishery processing waste
- Safe products with utility of optimization

Alternative Feed Ingredients

The questions posed to this group included:

1. What are potential alternative sources of protein and oil to fish meal and fish oil in aquaculture feeds, and are there specific obstacles to their use in aquaculture feed?
2. What modifications and processes show the most promise for improving the nutritive value of existing aquaculture feedstuffs or developing new feedstuffs, and how close are these technologies to commercialization?
3. What technologies are commercially viable or just technologically feasible to produce a source of long chain omega-3 fatty acids?
4. Who should lead the development and evaluation of alternative diets for aquaculture–industry, universities, government, NGO's, or collaborations of these?

In addressing these questions, the group identified a number of metrics to consider itemized challenges and possible alternative feedstuffs and discussed who should be involved in the research of alternatives feedstuffs. The metrics were as follows:

Key Metrics Included

- Economics and logistics of alternative ingredients
- Processing characteristics of alternative ingredients
- Supply and supply efficiencies
- Contaminants in all ingredients
- Public perception and production acceptability
- Immunostimulants
- Fish health–probiotics
- Environmental impact
- Sustainability
- Carbon Footprint

The top priority feedstuffs identified to explore were (several tied for 1st, 2nd and 3rd place):

1 Poultry byproducts
1 Industrial and food byproducts, coproducts
1 Insects and insect meals (e.g. soldier fly)

2 Yeast-based proteins
2 Fishery processing byproducts
3 Micro algal meals–high DHA and proteins
3 Macro algae meals
3 Zooplankton

Future Directions for Feed Manufacturing

The questions that this group was charged with include:

1. What are the high priority needs for feed manufacturing and processing that might be predicted over the next 5 years, based on current agricultural trends, emerging technologies and availability of resources?
2. What can be done, and who should do it (industry, NGO's, government, universities, or a collaboration of these) to strengthen our ability to meet future challenges in manufacture of alternative feeds for aquaculture?
3. What is the most reasonable strategy for near term steps while considering the long-term perspective of ensuring economic viability, environmental quality, and human/animal health?

The research needs for development of future feed manufacturing were identified as follows:

- Fundamental biology
  - Bioavailability (input/output)
  - Nutrient requirements
  - Artisanal aquaculture (niche, organic)
- Selection of species (for nutritional value)
- Engineering and technology
  - System constraints
  - Methods of extraction of nutrients o Processing technologies
  - Standardized/cost-effective toxicity testing
  - Energy inputs, uses and sources
- Market/production
  - Safety/surveillance
  - Coordination
  - Feed conversion efficiency and waste considerations
  - Regulatory challenges

- Societal needs
  - Transparency-data research results
  - Perceptions
  - Pilot demonstration
- New generation raw materials
  - Existing (e.g. bycatch)
  - Emerging

As to who should be supporting, conducting research and communicating results, the group identified the following recommendations:

- All research to be addressed by private industry with government support
- Third party verification of study results
- Research needs to be done by "neutral" organizations
- Government needs to be involved in ingredient-prospecting for novel & non-traditional sources
- Government should support nutrient requirements and update (NRC'93)
- Collaborative work public—private & government academia
- Government should take on selective breeding
- Government should be involved in all the research and disseminate to all
- NGO should support sustainable research on alternate ingredients
- NGO's should aid in education of public acceptance of alternative ingredients

At the end of the discussions and breakout groups, all attendees were asked to complete the following "homework" designed to provide a set of visions for the future of feeds for aquaculture:

> The Future of Aquafeeds . . .
>
> This is a take home assignment – each participant should send in within two weeks following the meeting, what they see happening in the next 5 and 25 years in the area of feeds for aquaculture. This is an exercise in science fiction so please take your best guess and use your imagination but be honest in what you really see as the future of aquafeeds. Please keep each Scenario (5 years from now and 25 years from now) to under 2 pages in length. As much as possible make them applicable to your location and species. Let us know what the diets will

be composed of, what the feed efficiency and growth rates will be and what breakthroughs occur to make your scenarios possible. Where will the limiting nutrients come from and what feedstuffs will dominate the industry in your country? What species will these diets be fed to? How much aquafeed is being produced worldwide? How are these diets sustainable in the long run? You are welcome to also put in natural disasters which might affect aquafeeds.

These "futurecasts" are presented in the next section.

FUTURECASTS FROM EXPERTS PANELS

The convening of both the scientific panel and the stakeholders panel provided the opportunity to gather opinions on how aquaculture feeds, aquaculture products and human nutrition will develop in the future. Participants in each panel were asked to predict the challenges and changes that aquaculture will face, and the developments that will affect both producers and consumers, over the next 5 and 25 years.

Summary

As noted before, at the end of the discussions, each panel member was assigned the following "homework" designed to provide a set of visions for the future of feeds for aquaculture:

> The Future of Aquafeeds . . .
>
> This is a take home assignment–each participant should send in within two weeks following the meeting, what they see happening in the next 5 and 25 years in the area of feeds for aquaculture. This is an exercise in science fiction so please take your best guess and use your imagination but be honest in what you really see as the future of aquafeeds. Please keep each Scenario (5 years from now and 25 years from now) to under 2 pages in length. As much as possible make them applicable to your location and species. Let us know what the diets will be composed of, what the feed efficiency and growth rates will be and what breakthroughs occur to make your scenarios possible. Where will the limiting nutrients come from and what feedstuffs will dominate the industry in your country? What species will these diets be fed to? How much aquafeed is being produced worldwide? How are these diets

sustainable in the long run? You are welcome to also put in natural disasters which might affect aquafeeds.

All the submitted future casts are collected in this section following the summary. Participants in each panel varied widely in background and expertise and represented multiple viewpoints. *Futurecasts represent the opinion and creativity of the authors, and should be considered works of fiction.*

Aquaculture Feedstuffs (Proteins and Oils)

Many contributors cited the scale of the biofuels industry as the factor most difficult to predict, and likely to have a major impact on feed composition. This industry will have a major effect on the prices of grains and all other related foods. Production of biofuels from grains will also generate tremendous quantities of coproducts such as distiller dried grains with solubles (DDGS). This product is currently of low value to the aquafeeds industry because of the low nutrient density, poor processing characteristics and variable quality standards, but in the next 5 years the quality of the product from some manufacturers will improve. In the next 25 years enhanced extraction and processing of DDGS will result in several types of products ranging from traditional DDGS to highly concentrated and digestible products. These products will be competitively priced and will likely be a part of all aquafeeds At the same time, the use of whole ground grains will decrease dramatically as the value chain for each crop changes, and the oil, carbohydrate and protein parts of the seed all have value. Necessity forces more complete use of each crop.

The wild harvest and processing of fish for human consumption produces tremendous quantities of byproduct in the form of offal from processing plants. During the pollock and salmon fishing seasons in Alaska for example, huge quantities of this waste stream are available in a short time window. Development of technology and capacity to stabilize the raw material for future conversion to feed will make this underutilized resource accessible and make fishing and feed production more cost effective. The resulting stabilized material will be a high quality protein and oil source for aquaculture feeds. Use of these products is expected to increase over the next 5 years with development of both on-shore and floating processing plants. It is predicted by some that within the next 25 years there will be full recovery and utilization of both commercial and sport fishing waste. It is also expected that marine harvest of new species such as urchins, isopods, and amphipods for inclusion in animal feed production will begin. It is reasoned that a far smaller proportion of the existing populations would be needed for feed if the harvest of animal protein is focused at lower trophic levels than the current anchovy, sardine and menhaden fish meal fisheries and the environmental impact would be far less. The harvest of krill is, and may continue to be, very controversial.

Current research demonstrates the feasibility of using a wide variety of non-traditional ingredients in fish feeds. These ingredients share a common approach of recycling nutrients, or capturing waste streams from other processes. The capture fishery by-products described above are a good example and others on the horizon include; micro-algae grown for feed using power plant CO off-gas as way to capture carbon dioxide and possibly coproduce ethanol and/or biodiesel from the algae; bacterial biomass produced by cleaning food processing wastes; fungal fermentations of plant materials to remove fiber, concentrate protein and remove anti-nutrients; insect larvae grown on a mix of substrates such as food wastes, DDGS from ethanol production, and livestock manure. Some of these ingredients sound far fetched at this time, but by the year 2033 it will be important to utilize every available nutrient stream, so that each of these products may play an important role in aquafeeds.

Long chain unsaturated fatty acids are one key to the heart and brain healthful properties of fish, and this nutrient class is currently the most limited of all nutrients. Currently, there appear to be limited options for obtaining the needed levels of EFA's. Genetic modification was one possibility identified. However, the controversy surrounding genetically modified organisms (GMO) has resulted in the aquaculture industry avoiding association with these products in any way. This is not thought to change over the next 5 or even 25 years for the animals that are produced. Nevertheless it is expected that feed ingredients that are produced from genetically modified plants will play an increasing role in the food supply, especially with regards to long chain omega 3 fatty acids. The demand for these important nutrients is exceeding supplies today, and there are few wild sources that can be harvested. These essential fatty acids are not naturally produced by terrestrial plants. In the next 25 years, oilseeds such as soybeans and canola will be genetically modified to produced these essential fatty acids in order to meet the ever expanding needs of the growing population. These same approaches may be used to add other nutrients such as vitamins or taurine to the plants to increase their nutritional value for both animal feed and human consumption. Genetic modification, mutation or selection will also be used to remove the anti-nutrients found in plants that are detrimental to animals and humans alike.

Aquaculture Animals

Changes in the cultured aquatic animal populations, to favor animals that can better utilize plant based nutrients and perhaps tolerate the anti nutrients, are anticipated too. Current estimates suggest that less than 20% of worldwide aquaculture production comes from genetically improved populations, and many species still rely on wild populations for breeders. Over the next 5 and 25 years the proportion of aquaculture production coming from domesticated, selectively improved populations will grow considerably.

Aquaculture Nutrition Information

While new and improved ingredients will be vitally important for the aquaculture industry in the future, expanding our knowledge of nutrient availability will be essential for realizing the full value of new ingredients. In addition, knowledge of the nutrient requirements for each species at different stages of life is essential for efficient use of feeds. For some of the major commercial species this information is currently being used, but over the next 5 years more information will be available and applied to commercial production. In 25 years, feed manufacturers will be able to match the nutrient requirements for all commercially produced species to the available nutrients in the feeds. Ingredient substitutions will be possible without reductions in fish performance.

Impacts on Human Nutrition

There is an increasing awareness from the public regarding the health benefits of eating fish. Consumption of the long chain omega 3 fatty acids found in coldwater fish has documented beneficial effects on health. Nevertheless, there is some sentiment that farmed fish have lower nutritional value than wild fish. It is thought that the wild fish have more long chain omega 3 fatty acids. The fact is that these fats come to the fish through the feed and high omega 3 fat levels can and are being achieved in aquacultured animals. A second misconception is that farmed fish may have higher exposure to environmental contaminants. The fact is that exposure depends strongly on their feed and their rearing environment. Diets of farmed fish are under increasing scrutiny and testing to assure the absence of contaminants.

The public is becoming increasingly motivated by the quality of food, and aquaculture products will continue to be improved over the next 5 and 25 years. It is expected that within the next 5 years, research on human nutrition will establish minimum, uniform recommendations on EFA's intake as it relates to human health.

In Conclusion

One condition was common to all scenarios and future-casts; human population growth continues and the demand for more fish and seafood increases. The need for aquaculture research to develop sustainable sources of nutrients and develop farmed aquatic animal populations that can thrive on a variety of nutrient sources is vital.

Individual Futurecasts

Futurecasts are in random order and are unedited. *Futurecasts are works of fiction and are limited to the imagination of the individual writers.* Futurecasts in no way represent any agency or groups official position. They were submitted by individuals who have an interest great enough in the topic to cause them to put in the effort to write them. The points we asked the authors to cover include the following:

- Please keep each scenario (5 years from now and 25 years from now) to under 2 pages in length.
- As much as possible, make them applicable to your location and species of interest.
- Let us know what the diets will be composed of, what the feed efficiency and growth rates will be, and what breakthroughs occur to make your scenarios possible.
- Where will the limiting nutrients come from and what feedstuffs will dominate the industry in the US?
- What species will these diets be feed to?
- How much aquafeed is being produced world-wide?
- How are these diets sustainable in the long run?

You are welcome to also put in natural disasters which might affect aquafeeds.

Jay Whelan
Department of Nutrition
University of Tennessee

What do we see happening in the next 5 years with aquaculture as it relates to human health?

The public will become increasingly concerned about their health. We have an aging population and neurodegenerative diseases are becoming more and more prevalent (real of perceived). Other health issues related to chronic diseases that have a relationship to nutrition will become an increasing concern. How can diet maintain health and how can diet treat existing conditions. Diet has an advantage over targeted therapeutic agents because of minimal side effects. Long chain omega-3 polyunsaturated fatty acids (LC n-3 PUFA) posses multiple action in maintaining health and treating disease. Intake is inversely associated with neurodegenerative diseases (dementia, Alzheimer's), coronary heart disease, sudden death, cancers, inflammation, etc. As we age, the risk of developing these chronic diseases will increase and the desire for non-invasive nutritional treatments will also increase. Critical to the aquaculture industry is their ability to address the major concerns of the public. The biggest is providing a safe product with consistent quality control. To do this standardization within the industry is critical.

Currently, fish derived from aquaculture is perceived to be inferior to wild fish. There are concerns about contaminants (pesticide runoff, etc) and about nutritional quality. The public has the perception that wild fish have higher levels of LC n-3 PUFA and may have lower levels of methyl mercury, dioxin, PCBs etc. The industry has to convince the public that the safest fish is derived from aquaculture and its nutritional and sensory qualities rival that of feral fish. The human researchers will be doing their part in establishing dose and response data needed to make informed recommendations. I suspect, in the next 5 years, their will be a drive to establish minimum, uniform, recommendations on n-3 PUFA intake as it relates to human health.

The demand for fish will increase if the concerns of the public are satisfied. I will predict that industry and governmental agencies will try to tighten quality control aspects. They will invest more money into research to establish these guidelines. The dietary requirements of different fish species need to be established through research that is published in peer-reviewed scientific journals.

The use of GMO feeds for fish will be a big issue and will need to be addressed through scientific investigation to minimize public concern. Genetic manipulation of plants that are used for fish feed will parallel the need for alternative food sources (i.e., plant meal and plant oils versus fish meal and fish oils). The technology exists.

What do we see happening in the next 25 years with aquaculture as it relates to human health?

Unless fish becomes a huge health concern, I predict that the need and use of fish products will increase over the next 25 years and this growth should be sustainable. The concerns over the use of biotechnical feeds will eventually subside due to market pressures and research that clearly establishes that the use of these types of feed have no human health concerns (if in fact that ends up being the case). Even now, the government's report about the health effects of LC n-3 PUFA asserts that the benefit of consuming fish out weigh the risks. Now, whether the global supplies can keep up with the public's demand is an issue that needs to be addressed by the industry. From a human health perspective, the consumption of fish and fish oils have shown and will continue to show health advantages over the use of other meats in the diet.

What are the long term solutions to existing barriers?

Research is the key. Research proper nutrition for fish and establish a database outlining these standards with proper testing to confirm the recommendations. Research the health effects in humans, including establishing the potential health benefits of "all" the nutrients in fish and fish oils (those nutrients in addition to LC n-3 PUFA). Genetically modify plants so they are adequate substitutes for fish meal and fish oil by enhancing required nutrients and minimizing anti-nutrients.

Craig Sheppard,
University of Georgia Entomology

Gary Burtle,
University of Georgia Animal and Dairy Science

Larry Newton,
University of Georgia Animal and Dairy Science

Available studies indicate that a complete or partial replacement of fish meal and fish oil with black soldier fly (*Hermetia illucens* L.) prepupae protein and fat will occur rapidly, in the face of decreasing fish meal supplies. This will be based on additional research on all aspects of production, processing and utilization of this novel feed.

Feed Efficiency and Growth

A replicated study indicated that complete replacement of fish meal in channel catfish diets is possible with no decrease in growth or FCR. In rainbow trout a 25 percent replacement of fish meal with prepupae meal gave similar growth. However, 50 percent replacement resulted in a significant reduction of growth, indicating a sensitivity to soldier fly prepupae meal at high rates of dietary inclusion. Both of these studies used whole milled prepupae. Separation of the cuticle (contains chitin) from the protein and fat of the prepupae will produce a superior feedstuff. A preliminary analysis indicated that cuticle weight is about 24 percent of the whole prepupae on a dry matter basis. Removal of this chitinous material will improve performance and increase the percent of fish meal that can be replaced in diets for sensitive aquatic animals as well as pay for the cost of removal by sales of the valuable chitin.

Production Breakthrough

Development of large scale production of black soldier fly prepupae will require scaling up and refinement of already proven systems. After large quantities of prepupae are available, byproducts from the production of soldier fly prepupae meal will include chitin, fat, protein concentrates and fine chemicals. Commercial equipment is, or will be available to separate these product streams and markets exist for the byproducts. By 2013, large scale production of soldier fly prepupae will be located in close proximity to the major aquaculture producing areas in the nation. The ability of soldier flies to use a variety of substrates for growth, while producing a consistent product, will makes them the preferred protein production alternative.

Species Fed, Designer and Organic Feed

Although plant protein has been shown to provide good growth for aquaculture species, including tilapias and catfish, soldier fly prepupae meal can replace fish meal in high efficiency diets and in the diets of carnivores. Dietary inclusion of up to 30 percent can be achieved for most fish and crustacean species. Development of a soldier fly prepupae meal after the chitin or skin has been removed will allow a higher level of dietary inclusion for sensitive aquaculture species, like rainbow trout. By 2033, all aquaculture diets will have soldier fly prepupae inclusions levels between 5 and 75 percent. Larval fish and crustacean diets will also be greatly impacted by the development of soldier fly prepupae meal due to their higher requirement for animal protein.

"Designer" and Organic Feeds

Supplementation of substrates for production of "designer" prepupae is highly desirable to increase the amount of omega-3 fatty acids and adapt the final composition to specific needs of aquaculture species. Diets for brood stock, especially shrimp and marine finfish, will benefit from designer soldier fly prepupae containing high oil, high omega-3 concentrations. Designer soldier fly prepupae will not be GMO, and may use substrates of "organic" origin to supply the food stock to produce "organic" and enriched feed-grade protein concentrate. The unique and controlled nature of soldier fly prepupae production allows for manipulation of the final nutrient composition of the product. This aspect provides an advantage over many protein products with variable source and manufacturing characteristics that change product composition, including fish meal, poultry by-product meal, and meat-and-bone meal.

Feed Needed

Aquaculture feeds that include commercial sources of fish meal, estimated on the basis of shrimp, salmonid, catfish and tilapia production, are about seven million tons worldwide. Base on the SOFIA projections for growth in the industry, 8.7 million tons will be needed in 2013 and 13.5 million tons in 2033. The amount of soldier fly prepupae meal needed will depend on the dietary inclusion rate and the rate of penetration of this new product into the feed protein market. The projected 10 percent average dietary inclusion rate and 50 percent penetration rate, 675,000 tons of soldier fly prepupae meal and protein concentrate will be needed by 2033.

Production Possible Prepupae production from just two previously researched "waste" streams from the U.S. alone can meet the projected needs. Also, many other feedstock streams and those outside the U.S. can be used. Production from manure of confined swine in the U.S. alone would be 1.8 million tons of prepupae. Another more easily accessed feedstock is U.S. brewer grains, which would produce about 500,000 tons of prepupae annually.

Insect Utilization, Nutrients, Antimicrobial

Many populations of wild animals, especially fish, birds, bats, and others, depend upon consumption of insects for a significant part of their nutrition. Most scientific effort devoted to insects has been aimed at destroying pest species, rather than using beneficial species. This vast production potential of insects will finally be used in support of the human food chain by developing the commercial production of *Hermetia* to feed our farmed fish. The composition of whole unseparated *Hermetia* prepupae meal is about 42 percent protein, 35 percent oil, 5 percent calcium, 1.5 percent phosphorus, 3.4 percent lysine, and 1 percent methionine/cystine. Interestingly, *Hermetia* meal oils contain about 54 percent lauric acid which has been shown to be active against oil coated viruses, including HIV virus, measles virus, clostridium, and many pathogenic protozoa. The high quality and unique characteristics of *Hermetia* meal will have many uses in the aquaculture and other animal industries.

Sustainability

Black soldier fly (*Hermetia*) upgrade lower value materials into a source of animal protein (and oil) which has a much higher value, especially as an ingredient in diets for fish. When these lower value resources (such as waste food, by-products of the alcoholic beverage or ethanol industry, or animal manure) are consumed by Hermetia, the process also serves as a treatment that reduces the volume and nutrient content of the starting material. This is especially important, for increasing sustainability, in the case of animal manures (which are often produced in excess of local demand) and materials that would otherwise end up in landfills. *Hermetia* production will be sustainable in that it uses underutilized materials and, in turn, it increases the sustainability of the endeavors producing these materials. The demands of the U.S. economic system which favors large scale production, established supply chains, and centralized distribution; puts distributed production (which is often more sustainable and usually the place of new commodities), at a distinct disadvantage. Systems of distributed production usually develop to supply a specialty market. The aquaculture feed industry will be enough of a specialty market to support the development of *Hermetia* culture into commodity status within 5–10 years. Other animal industries, that use fish meal, will also benefit from the availability of *Hermetia* meal. With sufficient penetration of the feed protein market by *Hermetia* meal, the demand for wild capture fish meal will move more into line with sustainable harvest.

Jeff Silverstein
USDA/Agricultural Research Service

In 2008

Protein-heavy reliance on fish meal, supply is limited. Protein alternatives exist, research ongoing to develop alternatives, examine impacts on product quality.

Oils-heavy reliance on fish oils, supply is severely limited. Oil alternatives exist for some of the demand, but HUFA (esp. EPA/DHA) alternatives not certain. Some precursors available from transgenic plants (stearidonic acid?)

Feeding-phase feeding and definition of requirements for different phases, endpoints is in early stages (at best).

Genetic improvement of fish, very early for all but Atlantic salmon (24–28 months to harvest @ 4kg) and maybe rainbow trout (9 months to harvest @500g). Even the salmonids have great potential for improvements in growth, possibly in broadening the materials that can be used/tolerated in feeds. Genetically improved populations supply ~15 percent of aquaculture production.

In 2013

Protein-Variety of protein sources in use and phase feeding important to minimize problems with specific compounds

Oils-the critical limitation, focus on phase feeding to meet requirement pre-harvest, and meet consumer need at harvest. Discussions about use of transgenic plants to deliver the HUFAs, transgenic animals to make their own HUFAs.

Genetic improvement progressed for salmonids. Harvest salmon at 20–24 months, rainbow trout at 8 months. Genetically improved populations supply 30 percent of aquaculture production.

In 2033

Protein-dietary protein will be supplied from a variety of ingredients, aquaculture will still be a big user of fish meal, but a smaller and smaller fraction per kilo of diet (just more kilos of diet). Delivery, absorption and availability of amino acids will be the major issue-having all the a.a.'s available at the same/right timing. All requirements for amino acids for the major species will be known and met for up to 5 separate growth intervals.

'Special factors' in fish meal will be identified.

Oils–New sources of HUFAs available from algae, possibly transgenic sources of HUFAs. The requirements for various species of FA will be defined (e.g. the roles of EPA/DHA). Phase feeding will be routine.

Krill will be harvested sustainably and provide one alternative for marine oils and proteins.

Genetic improvement-fish will supply a much bigger share of proteins in the world food supply. Genetically improved populations will supply the majority of aquatic foods. Growth rates will be the prime improvement, followed by disease resistance. Salmon will be harvested within 6 months of seawater transfer. Trout will be harvested 5 months from hatch. Larval feeds will be optimized, selection on artemia and copepods for smaller sizes to feed first feeding juveniles.

Rick Barrows
USDA–Agricultural Research Service

Past and Present

Research on replacements for fish meal in aquaculture diets has been conducted since the early 1970's. At that time fish meal and fish oil was relatively cheap and supplies were large and high quality. The primary factor driving the early research on alternative protein sources was simply to reduce feeds costs. In the late 1990's research intensified on a global scale to find replacements for fish meal. Many factors were driving this research effort including increased demand with static supplies, a more restrictive regulatory environment, and detection of contaminants in fish oil and meals. Recent studies have demonstrated that fish meal can be totally removed from the diet of some carnivorous species without a reduction in fish performance. This research demonstrates that elimination of fish meal from carnivorous fish diets is feasible, but other obstacles remain for other species.

The problem facing the aquaculture feeds industry today is not just finding better ingredients. The situation is more complex with several major factors limiting progress in removal of fish meal from aquaculture feeds. First, relative to terrestrial animal species little is known of the complete nutrient requirements for the variety of aquaculture species. Some amino acid requirements have been determined for several species using purified diets, but this information does not seem to be effective with practical feeds fed to fast growing strains of fish. Second, specific nutrients not normally considered when formulating fish meal based diets must be accounted for in the diet. Some of these nutrients include, taurine, hydroxyl-proline, macro-minerals, and insositol and more are currently being identified. Third, the number of ingredients available with the proper nutrient composition (i.e. protein greater than 55-60 percent) is very limiting. Alternative sources of essential fatty acids (other than fish oil), both for fish metabolism and human health, are currently not available. Fourth, the industry needs to move away from formula based diets and into nutrient based diets, but a better understanding of the availability of nutrients from the alternate ingredients, and the requirements of the fish, is critically needed for this progress to happen.

Five Years from Now

More of the essential nutrients in fish meal have been identified and are supplemented or balanced allowing increased use of alternative ingredients. The role of plant products in aquafeeds will be strongly influenced of the expansion of the biofuels industry. As ethanol production consumes more grains, methods to improve the quality of DDGS, resulting in lower fiber products with increased amino acid availability will be developed. A whole series of protein ingredients will be developed on the basis of sustainability and nutrient recycling, and will be evaluated with many species, but not yet comprising a major portion of the aquafeed market. These include micro-algae's grown using power plant off-gas as way to capture carbon and possibly produce ethanol and/or biodiesel at the same time. Fishery processing wastes will play an increasing role as traditional capture fishery meal replacement, but primarily in starter feeds. Bacterial cultures that are used to clean up food processing waste streams will then be used in aquafeeds. Fungal fermentations will be used to consume anti-nutrients from plant products and produce proteins. Insect larvae grown on a variety of coproducts including DDGS from ethanol production, food wastes, and livestock manure will be used in some feeds. Phase feeding fish oil (used only during the last phase of production) will be common practice to conserve the limited supplies. At the same time more research on human nutrition will help in determining the target levels for essential fatty acids in fish fillets.

Twenty Five Years from Now

All species of fish produced in large scale will be using nutrient based formulations to allow for fluid transitions of ingredients in the diet as ingredient cost and availabilities change. Nutrient requirements will be known for the major life stages for many species. Out of necessity, and due to rigorous testing, the public has accepted the production of genetically modified oilseed crops (soybeans, canola, flax) that contain the omega 3 fatty acids previously found only in fish oil. If these products are affordable they will be fed throughout the life cycle of fish, or perhaps just used as in finishing diets as fish oil was used in the past. The majority of fish oil will be used directly as human food. Modified ingredients tailored for specific aquaculture feeds will be fully developed and utilized in aquafeeds. Nutrient recycling is not only important, but the only economically way to continue production. Modification of plant products through processing or genetically will remove the problem of anti-nutrients. Enhanced strains of fish will improve production efficiencies and remove any problems with palatability of alternative ingredients. In summary, better fish, more alternative ingredients and improved feeds will result in a strong industry providing a healthful product to the consumer.

Murray Drew
Department of Animal and Poultry Science
University of Saskatchewan

Assumptions

Worldwide aquaculture production has grown at 10 percent per year and has tripled to approximately 100 million metric tons per year. Assuming an overall feed conversion of 1.5 this means 150 million metric tons of aquafeeds per year.

Fish meal and oil production are static at 6 million metric tons and 700 thousand metric tons respectively. Assuming 100 percent of these products were used in aquafeeds that means 4 percent fish meal and 0.47% fish oil in the average aquafeed formulation.

We will need approximately 10 million metric tons of high quality protein and 2 million metric tons of vegetable oil containing high levels of highly unsaturated fatty acids to replace marine products in aquafeeds.

Protein

Anti-nutritional factors (ANFs) present in plant proteins mean that we will have to develop: 1) biotechnology produced crops with lower levels of ANFs than current crops and 2) methods of processing crops to reduce levels of ANFs.

The palatability of plant proteins is lower than that of marine products. This will require: 1) the development of palatability enhancers and 2) strains of fish that accept high dietary levels of plant proteins.

Complex diets formulated with many plant proteins support higher growth performance than simple diets. We will require a moderate number (more than 6) of high protein ingredients for tomorrow's aquafeeds.

Increase the amount of marine products available for aquaculture by harvesting products at a lower trophic level.

Oil

Fish oil contains high levels of highly unsaturated fatty acids (HUFAs). To maintain the nutritional quality of fish products, alternatives to fish oil must also contain HUFAs. This will require biotechnology produced oilseed crops that produce HUFAs.

Products We Will Be Using in 2020

- Aquaculture grade soybean protein concentrate (Low ANF varieties processed to increase protein content)
- Corn gluten meal (an old standard)
- Canola protein concentrate (60 percent crude protein with low levels of ANFs)
- Pea protein concentrate (55 percent crude protein with low levels of ANFs)
- Distillers dried grains and solubles protein concentrate (relatively cheap and palatable)
- Aquaculture grade marine meal (harvested from low trophic level marine organisms)
- Fish meal (used at low levels to improve palatability and growth performance)
- High (HUFA) soybean and canola oil
- Poultry meal
- Fish oil (used in final finishing diets for flavor enhancement)
- Palatability enhancer (plant based protein source with attractant/ palatability enhancer properties)

Areas of Required Research

Plant science-development of crops specifically for aquaculture diets.
Soybean –proteins and oils
Canola-proteins and oils

Aquaculture nutrition
Develop a better understanding of the effect of plant ANFs on fish metabolism, physiology and intestinal microflora.
Use of new products in aquaculture diets.

Fish nutrigenetics
Develop improved strains of fish that better tolerate plant proteins and oils in diets.

Human nutrition

Increased knowledge of the benefits of fish in human diets is essential to developing consumer buy-in.

Marketing

Improve the acceptance of farmed fish by the consumer.

Both nutritional and environmental aspects of aquaculture products must be promoted.

Takao Yoshimatsu
Fisheries Research Agency (FRA), Japan
National Research Institute of Aquaculture (NRIA)
Feed Research Group, Aquaculture Systems Division

Background and Present Situation

Japan used to be the leading fish meal manufacturer in the world. For example, in 1980s there were constant 3-4 million mt or more of sardine (*Sardinops melanostictus*) fisheries production from the surrounding seas of Japan. At that time this huge catch of sardine substantially supported the large production of developing sector of the aquaculture of high commercial value fish like yellowtail (*Seriola quinqueradiata*) and red sea bream (*Pagrus major*). However after 1990s Japanese sardine production drastically decreased mainly due to the influence of natural regime shift occurred in oceanic environment with long-term intervals. To compensate this collapse of sardine production Japan started to import large amount of fish meal products from Peru, Chile and many other countries, and its total amount reaches 400–500 thousands mt nowadays. Approximately 50 percent of the imported fish meals are used as a feed ingredient for cultured fish and eventually 400–500 thousands mt of formula fish feeds is produced for finfish culture in Japan. However the recent soaring of fish meal price due to the increase in demand and international scrabble for fish meal products make the situation difficult from economical reason. In addition to that due to the influence of global climate change, the productions of pelagic feed material fish like sardine and anchovy are not always stable and predictable anymore. On the other hand, world aquaculture has become a dynamically developing sector of the food industry in recent years. With stagnating yields from capture fisheries and increasing demand for aquatic products, expectations for aquaculture to increase its contribution to the world's production of aquatic foods are very high, and there is also hope that aquaculture will continue to strengthen its role in contributing to support human health and sustainable food supply.

Under these state of affairs, for the advancement of aquaculture sector in Japan, the National Research Institute of Aquaculture has been participated an exceptionally imperative role as a governmental research institute hitherto. Its recent research activities on feed ingredients can be categorized as follows: 1) Efficient utilization of plant-derived feed ingredients like soybean meal, 2) Efficient utilization of plant oil and carbohydrate as feed ingredients for marine finfish, 3) Necessity of taurine as an important nutrient for marine finfish, 4) Improvement of fish feeds without fish meal using bile salts and taurocholic acid, 5) Efficient utilization of single-cell materials produced from seaweed biomass by enzymatic means for aquafeeds. The outputs from these research activities made a great contribution for the development and improvement of aquafeeds used for finfish culture not only in Japan but also all over the world. However we are now required to propose novel and innovative ideas to give breakthrough this difficult situation surrounding aquafeeds mentioned above. As the head of feed research group of NRIA I imagine and forecast my personal opinion on the future of aquaculture and aquafeeds in Japan in 5 and 25 years time period from now. To avoid confusion the subject of the following prospects restrict only on feeding cultures of finfish and aquafeeds used for them.

Five Years Later from Now

Net-cage culture of highly-economic finfish will be amazingly popularized in Japan. The main target species are marine species like blue-fin tuna (*Thunnus thynnus*), amberjack (*Seriola dumerili*), yellow tail (*Seriola quinqueradiata*), red sea bream (*Pagrus major*), tiger puffer (*Takifugu rubripes*), Japanese flounder (*Paralichthys olivaceus*), groupers (*Epinephelus* spp.), etc. moderately a huge quantity of flesh harvest will be exported to the new market formed in main terrain China and other Asian countries. The consumption of seafood will be increased among the prosperity living coastal area in China, and eating puffer fish is not illegal any more. For some species like red sea bream and Japanese flounder commercial hard pellet formula diets are given by using computer-controlled automatic feeders equipped with solar panel. Sophisticated demand-feeding apparatuses are developed and commonly utilized to minimize the loss of feed and unexpected load of N and P to the water environment. Some parts of fish oil can be successfully replaced by DHA extracted from cultured protist biomass like marine net slime molds *Labyrinthula*. More than 50% of dietary protein is plant-derived protein like

DDGS that are completely proved to be safe as human food and for environment.

For some important cultured species like blue-fin tuna, amberjack, yellowtail and tiger puffer, so-called Oregon-type moist pellet are still used for growing out to commercial size. This modified Oregon-type moist pellet are comprised of fish process by-products, single-cell materials made from seaweeds and local fish meal made from Pacific saury (*Cololabis saira*) and anchovy (*Engraulis japonica*). Harvested fish are processed into fillet and/or block forms near aquaculture sites and ship to the market or other area to reduce the transportation cost. Fish trimmings and left-over wastes can be reused as one of the main feed ingredients without freezing storage; i.e. lighter eco-mileage and eco-friendly feedstuff. Near the net cages for fish culture aquafarmers are encourage to culture edible seaweeds like laver (*Porphyra* spp.), wakame (*Undaria pinnatifida*), and sea tangle (*Laminaria japonica*) as water quality conditioners; a kind of new integrated fish culture. Good seaweeds products are supplied for human consumption and pharmaceutical use, and the other low grade products will be used as feed ingredients after some single-cell processing using enzymatic means. The import of high price fish meal from overseas might be gradually reduced due to the exploitations of new fish meal materials that can be harvested near the seas of Japan. In general, aquaculture would be progressively a growing sector of fisheries and its products are recognized as good commodities to earn money not only in Japan but also all over the world for a while.

Twenty Five Years Later from Now

From world wide point of view, the main role of aquaculture will be completely altered due to the explosive increase of world's population and the unexpected change of global climate in future. According to the many statistical data the world's population, on its current growth curve, is expected to reach nearly 9 billion by the year 2050. This number is certainly beyond the limit that the earth can support its food for world people to live on. In 2033, 25 years later from now, the situation surrounding us will be getting closer to this critical phase. Under this difficult condition the importance of aquaculture would become greater and greater to produce good food resource for humans. People cannot depend on small pelagic fish as feed ingredients for culturing fish anymore. Most of the fish caught in the sea are used as excellent foods for human being; even they are very small in size. Almost all the feed stuff for aquaculture will come from organic by-products and wastes. By utilizing DDGS obtained after biofuel processing as one of the main protein source omnivorous and herbivorous low-trophic level fish which require less dietary protein will be widely cultured and supplied to the consumers as cheap and healthy foods.

For culturing high commercial value fish, most of them are highly carnivorous, fish wastes and meals from terrestrial animal wastes are used as main feed ingredients. For CO2 fixation genetically modified high-performance microalgae are widely cultured in ponds near factories and power plants. By feeding these microalgae huge amount of microscopic plankton (*Brachionus rotifers*) can be produced. Also as rotifers are utilized for sewage treatment people have continuous supply of this nutritious zooplankton. After extracted oil for biodiesel, rotifer wastes are utilized as alternative protein source for feeds for fish as well. Of course this rotifer is a GM strain which produce enough amounts of HUFAs including DHA and EPA.

Genetic modification technology and GMO will be played a very important role when we use crop meals and DDGS as a main feed ingredient by modifying its nutritional profile and improve its nutritional faults and demerits. Nevertheless we should always keep it in mind that it is prerequisite to be proved and insure safety as human foods and for environment from any view points.

References

Kalla, A; Yoshimatsu, T; Araki, T; Zhang, DM; Yamamoto, T, Sakamoto, S. Utilization of Porphyra spheroplasts as a feed additive for red sea bream. *Fisheries Science*, 74, 104-108 (2008)

Khan, ND; Yoshimatsu, T; Kalla, A; Araki, T; Sakamoto, S. Supplemental effect of Porphyra spheroplasts on the growth and feed utilization of black sea bream. *Fisheries Science*, 74, 397–404 (2008)

Yoshimura, K; Usuki, K; Yoshimatsu, T; Kitajima, C., Hagiwara, A. Recent development of high density mass culture of the rotifer Brachionus rotundiformis. *Hydrobiologia*, 358, 139-144 (1997)

Yoshimura, K; Tanaka, K; Yoshimatsu, T. A novel culture system for the ultra-high-density production of the rotifer, Brachionus rotundiformis—a preliminary report. *Aquaculture*, 227, 165-172 (2003)

Yoshimatsu, T; Kalla, A; Araki, T; Zhang, DM; Sakamoto, S. A preliminary report on the use of Porphyra protoplasts as a live food substitute for culturing aquatic animals. In "Aquaculture and stock enhancement of finfish– Proceedings of the thirty-fourth U.S.-Japan Aquaculture Panel Symposium (ed. by Stickney R, Iwamoto R, and Rust M.), NOAA Technical Memorandum NMFS-F/SPO-85, 63-67 (2007)

Tsuyoshi Sugita
Fisheries Research Agency (FRA), Japan
National Research Institute of Aquaculture (NRIA) Feed Group, Farming Systems Division

5 Years from Now

In Japan aquaculture industry of 5 years later, the economic and environmental problems from fish meal probably has become more serious, and utilization of the alternative protein sources will be promoted. Plant proteins such as defatted soybean meal and corn gluten meal are practical alternative protein source. As reported at the meeting of the other day, taurine is very important for marine fish species, which are main aquaculture species in Japan. However, the plant protein sources are not containing taurine. There are 2 types of nature and synthesis in the taurine. Since the price of natural taurine is high, the natural taurine cannot use as a feed additive. Although the synthetic taurine is very reasonable price, the use of synthetic taurine as a feed additive has not yet authorized by the Japanese government. Therefore 5 years from now, the Japanese researchers will be made a several attempts to obtain the judicial applications of synthetic taurine use as a feed additive.

Presently, in Japan the use of meat and bone meal from bovine as a feed stuff is under ban of the law for the influence of bovine spongiform encephalopathu (BSE). Furthermore, for the influence of avian influenza the use of chicken meal as a feed stuff is sidestepping from the consumer. Therefore, the appearance of novel animal protein source is expected. Soon, the use of meat and born meal from pig as a feed stuff will be authorized by the Japanese government. Therefore 5 years from now, the Japanese researchers together with feed companies probably examine the practicality of meat and born meal from pig.

25 Years from Now

It is very difficult to expect the future of 25 years later. When present condition is considered, fish meal and fish oil probably could not use as feed stuffs in 2033. As reported at the meeting of the other day, the gene recombined DHA and EPA containing soybean and phytate free soybean will be authorized by the United States. Furthermore, the gene recombination will probably make the optimal soybean for aquafeeds (e.g., lysine, methionine and taurine rich soybean, free from antinutritional factors (i.e., lectines, antigenic and estrogenic flavons, oligosaccharides, and else) free soybean. Presently, Japanese many people have negative image in the gene recombined foods.

However if United States promotes the utilization of gene recombined soybean, Japan cannot avoid using that soybean, so Japan has imported much soybean from United States. Therefore, in order to improve the negative image for the gene recombined soybean of the Japanese people, the Japanese researchers will examine the safety of gene recombined plant source.

Roberto Marchesini

As the note you gave us said: "take your best guess and use your imagination", I took some liberty to write down some ideas I have and imagined what I would like or what I might think will happen to aquaculture and aquaculture feeds. I hope that you like what you read and also that this will help you to do your work.

I was a foreign participant at the meeting, I live and work in El Salvador, Central America as an aquaculture consultant and I work mainly with tilapia, freshwater and marine shrimp in semi-intensive systems. This exercise about aquaculture and aquacultured feeds problems has taught me a new tool that I might use to propose the development of aquaculture policies in my country.

I would like to thank you and the organizers of the event for letting me attend and express some of my ideas. Please send them my regards and congratulations.

Sincerely yours,
MSc VMD Roberto Marchesini

This Is the Year 2013 in USA

America is working harder on *Tilapia*, marine and fresh water shrimp culture. The diets formulations on those animals have been changing and by now those aquatic animals nutritional requirements are known. Essential fatty acids oils of fats and essential amino acids of the proteins minimum percentages of the diets on those animals on their different stages of their lives have been researched and revealed by the National Research Council, the most important agricultural oriented universities or other scientists. Those discoveries have been accomplished thanks to the available grants given by the Sea Grant and some other governmental entities to the interested scientific community in the following of this line of investigation.

Aquaculture of *Tilapia*, fresh and marine shrimp is now more intensive due to the public demand of those products. Due to the specificity of the nutrient requirements on the diets there are needs of cheap sources of fat and proteins (essential fatty acids and amino acids, that is) and so there is an emerging industry of guts and leftovers recycling of the main fresh and seawater fishing and aquaculture industries (salmon, Alaskan crab, trout and some others).

Feed conversions and growth rate still remains almost as 5 years before without much of a change, but aqua-farmers income is somehow increased not because of the reductions of the cost of the main feed ingredients (fats and proteins) but because of the appropriate nutrition of the broodstocks. Price reduction of feeds is not quite a reality yet because of the increase in prices of the other ingredients of the feed, for instance corn, sorghum, gluten and wheat meals have sky rocketed because the Biofuels policies in the US and worldwide.

Those diets will work for now but aquaculture has kept on growing at a steady pace for the last decades that there might be a shortage over the next years on the quality and quantity of the feeds ingredients.

More to come....

This Is the Year 2023

Industrial intensive and super-intensive aquaculture is the way America has chosen to compete against cheap, polluted, non-environmental friendly seafood raised in China and other eastern countries.

Freshwater and marine shrimp and Tilapia have been under a protocol of genetic improvement for a few years now. Many of the nutrients probed for adequate human health and nutrition are fulfilled by aquacultured seafood. Essential amino acids and fatty acids needed in children, mature and elderly people are commonly found on seafood, because those aquatic animals have acquired capability of retain certain nutrients of feed.

By now feeds in many places are a special mixture of certain strains of intensive cultured bacteria, microalgae and yeast. Following the lead of Norwegian scientists, US industries developed bacterial cultures and genetically modified them to obtain all components of the different nutrients needed in aquaculture feeds. Those microorganisms are nurtured on LPG (like the Norwegian strains) or a mixture of sunlight and special nutrients in the form of fertilizers.

Microalgae, bacteria and yeasts now easily produce fatty acid oils and amino acids in intensive facilities, those are required by aquaculture feed industries in the form of semi-liquid biomass and are processed to produce dry protein meals, fluid essential oils and unspecific growth and immunologycal stimulant factors. Nutritional requirements of aquacultured broodstocks are known since many years ago and those environmental friendly cultures are supplying what is needed.

By now feed efficiency is almost reaching 1:1 levels and growth cycles have diminished around 15 percent compared to 25 years ago because of the high digestibility of feeds and the unique improvements in genetic lines of broodstock.

Apparently the major limiting feed ingredients are corn, wheat and sorghum meals and similar due to environmental factors originated by global warming that has affected crops with droughts, hurricanes and related events.

Anonymous One

In looking forward to 2013 and 2033, it is apparent that the continued growth of aquaculture will demand more and more sources of natural proteins. This will be based on the international interests in "lifestyles of health and sustainability" (LOHAS), the interest in natural and organic products, and the awareness of many unknown health factors with "genetically modified" foods and from contaminates through chemical enhanced growth and processing. This will help spur a market for natural foods that can be made from unused biomass available in fish and invertebrates, including marine and land based organisms.

From Alaska, the future sources of aquaculture food materials exist in three areas: recovery of biomass from fish processed for human consumption, including salmon, pollock, cod, halibut, other flatfish, and rockfish; new reduction fisheries for arrowtooth flounder; and harvest of aggressive marine invertebrates, including urchins, isopods and amphipods for reduction.

While large biomasses exists in Bering Sea clams, arctic krill and other euphausiids, they are important food sources with ecological links through the food chain to marine mammals, also key elements to the Alaska's marine ecosystems.

In this context, I offer the following perspectives on the year 2013 and 2033:

In 2013, market forces, business interests, and regulatory schemes will move towards optimized use of the existing seafood processing biomass. This will build upon the present fish meal and oil recovery capacity in the shoreside plants and offshore catcher/processor vessels. Existing plants will be increased in capacity through a stabilized fish processing waste process, allowing peak periods to be carried through times of limited returns. New floating plants will be located in Prince William Sound and Bristol Bay to allow handling of the processing volumes in the surrounding area. The support tugs that brought the floating plants to the area will work in transporting bladders of stabilized waste from various fishing communities to the floating plants. Meals and oils will be stored in the nearest community awaiting shipment to a major cargo port upon completion of season. New shoreside plants will be planned in the Prince William Sound, Bristol Bay, and the upper areas of Southeast Alaska with particular interest in salmon, based on the business viability of the floating plants. As developed shoreside, floating plants will move to areas other areas as fishery advances. A sustained stream of high quality meals and oils will be provided to the domestic aquaculture feed manufacturers to insure contaminant and disease free food based on cold water marine resources.

By 2013, a new fishery will open for urchins as a natural marine source of color in the flesh of aquacultured fish. This, combined with the natural color in salmon meal and oil, will provide a more robust color to many of the aquacultured fish. Grants will be in place for research to find effective ways to: harvest isopods and amphipods allowing for a new source of aquaculture food while improving the quality of fish caught with pots and longlines; support the handling of sport charter caught fish processing waste and enable transport of the daily catch waste to be added to a stabilized waste stream; harvest Arrowtooth Flounder for reduction; develop appropriate food blends that maximize the use of marine and terrestrial plant and animal proteins, oils, and nutrients.

By 2033, Alaska will have full recovery of the commercial and sport fishery waste, as well as managing the aggressive marine invertebrates, thus improving the ecosystem. Alaska will venture away from export of meals and oils to domestic aquaculture to enhanced ecosystems with open-ocean ranching and specialized foods in a sustained large marine ecosystem approach. Alaska's continental shelf will be maintained in a balanced ecosystem with sustained productivity of preferred species. Specialized ships will distribute feed to lure preferred species into concentration at known locations for very efficient harvest (high cpue). Regional enhancement facilities will be growing fish, including halibut, sablefish, and pacific cod, to a post larval stage for return to the "open ocean ranch". Offshore fisheries will show comparable growth rates and production to that seen in the history of the Alaska salmon enhancement programs. A plateau will be reached with sustained TAC of 4.5 Million MT and a TAC on halibut, cod, and sablefish that is 3 times the 2007 harvest level.

In summary, the return of the natural fish and marine invertebrate proteins and oils to the ecosystem, along with an enhancement program for the preferred ocean fish of our region will lead to an enhanced ecosystem that optimizes the annual production of the large marine ecosystems of Alaska's continental shelf. However, to get there, Alaska will need to optimize the use of its marine resources by maximizing the recovery of seafood processing waste and conducting reduction fisheries on unused biomass, including Arrowtooth Flounder and the aggressive natural marine predators of urchins (herbivore on kelp habitat) and isopods and amphipods (captured fish). These materials will need to be used to nurture fish and establish growth rates on the important indigenous species, as well as develop feeding/ aggregating behavior schemes, so that open ocean ranching through ecosystem enhancement can be managed and the species effectively harvested. A national aquaculture program can provide just such a research and development tool to answer these questions in a controlled but commercially viable environment.

Anonymous Two

Five Years out

Work still being done to replace or decrease the use of fish meal and fish oil in aquafeeds.

A more efficient and cost effective way to clean fish oils of contaminants will be developed so more oil is available for feeds and human consumption.

Microarray analyses will have been used to determine the effect of new ingredients on the fish at a genetic level. This information will help guide the choices of ingredients to use in fish feeds. EPA will have a greater role in deciding how much of the nation's crops are converted to biofuels and how quickly this happens. EPA's decisions will affect the price of commodities including corn and soybeans.

Twenty-Five Years out

Ocean waters continue to see a drop in pH. The changing environment in the oceans will have a major impact on the food chains and productivity negatively impacting the remaining fish stocks; this includes both food fish and fish meal stocks.

Strains of fish have been selected for that can tolerate and thrive on high levels of soybean meal and other vegetable proteins in their diet. With the evolution of biofuels an "if then" scenario will occur. If the biofuels industry continues to grow then the byproducts of biofuel manufacture will be used in feeds. If biofuel development does not grow as much as expected then more grains will be available for feeds. Basically it is unknown how biofuels will impact the availability of feedstuffs.

Anonymous Three

Species known–salmonids, especially Atlantic salmon diets must include animal protein sources because salmonids are carnivores so need the sum of all of animal matter (amino acids, nucleotides, organic acids, taurine, methyl amines etc). Plant protein concentrates are expensive and must include too many micro-nutrients to be a viable replacement of fish meal. However, plant protein concentrates are extensively used and are important. Growth rates are influenced at least equally by farm management/environment and feed so feed can be made with cheap materials ie. fish meal can be replaced. FCR is critical and a good index to evaluate plant + animal protein mixes.

Many potential sources exist but economic sources are very rare and limited. Less necessary than "breakthroughs" is competition amongst animal feed protein suppliers to produce high quality and low priced feedstuffs. The industry is relatively small and will benefit most from new suppliers rather than new ingredients. This may mean that large food suppliers will be attracted to the aquafeed industry. In particular protein concentrates of 60+ percent are limited in supply.

Breakthroughs will be in better understanding of the negative nutrients. We know well the essential nutrients but each new product tested seems to not perform as predicted by amino acid composition due to "anti-nutrients". A public funded consortium in this area is critical.

The most limiting nutrient for salmonids will certainly be omega 3. The IFFO and FAO have projected that fish oil demand (the only current economic source of omega 3) already exceeds supply and prices are rising rapidly. In addition the human food industry is adding omega 3 to milk, yogurt etc and driving demand and prices up even more. The solution is micro-organism cultures and companies exist. That this product becomes economical so that consumers eat oil rich fish for health versus cheap chicken is a great challenge, perhaps requiring special targeted funding.

Protein sources can and should come from waste or coproducts or by-products. In particular poultry meal made from chicken processing waste is an excellent alternative to fish meal and sustainable.

Thus, sustainability is much more dependent on fish oil replacement for which there are no alternatives today.

In addition, El Nino years cause a 30+ percent drop in fish meal and oil and the next one (?2010) will be an economic disaster for the aquafeed industry.

Anonymous Four

Aquafeeds–2013

Biofuel byproducts/coproducts (primarily DDGS) continue to grow in quantity. By 2013, it is predicted that many ethanol manufacturing plants will implement very aggressive quality control programs, which will substantially reduce variability (e.g., in moisture content, total protein, amino acid profiles, etc.) for individual plants over time. Still, however, there are drastic differences between processing plants. Thus, use of these coproducts in aquafeeds will be dependent upon selecting DDGS from a specific processing plant that will appropriately meet the nutrient requirements for specific end-users.

Storage and handling (flowability issues) will be resolved, which makes their use much more favorable, regardless of location without, or even outside, the country (i.e., transportation and logistics is no longer an issue).

The quantity of DDGS produced will have grown to such an extent that DDGS has become a much lower-cost feed ingredient versus other vegetable-based materials. As such, use of DDGS will completely replace other corn-based materials which have historically been used in aquafeeds. The impetus for this change has been driven by the fact that more corn will be used for conversion to fuel ethanol vis-à-vis other end uses. At this point in time, there is not only great interest in using DDGS as an aquaculture feed ingredient, primarily as a supplementation versus a complete protein source, but many commercial feeding operations will have proven their effectiveness.

Many feeding trials will be conducted by 2013, in a variety of species. Promising results will be seen with inclusion levels of unmodified DDGS of up to 30 percent. High-protein (i.e., fractionated) DDGS can be effectively used up to 40 percent inclusion. A key to these high levels is the balancing act between providing appropriate nutrients (provided by the complete rations) versus optimizing processing techniques and conditions, in order to produce floating, pelleted feeds that do not have any binding problems.

Aquafeeds–2033

At this point in time, world oil production is in drastic decline, and the push for alternative fuels continues to escalate. World population growth, and its attendant repercussions, has nearly reached the earth's carrying capacity. The demand for food outstrips supplies, and aquaculture is increasingly relied upon to meet these needs.

Use of DDGS (both traditional/unmodified, as well as fractionated high-protein) are well established in aquafeeds by 2033. The biofuels industry has grown beyond the use of corn grain alone, however.

Biodiesel and lignocellulosic conversion processes have become economical and large scale. Unfortunately, many of the traditional feed ingredients are no longer available (for either livestock or aquafeeds), because they can be readily converted into biofuel as well.

Byproducts from these novel biofuel processes have grown substantially, and researchers must find ways to replace traditional feed materials. Preliminary feeding trials have shown that they have nutritional benefits for aquafeeds. Substantial research still needs to be conducted to determine which byproducts are appropriate for which species, and what inclusion rates are optimal for each.

In: The Future of Aquaculture Feeds
Editor: Ella Linton
ISBN: 978-1-63463-069-6

Chapter 2

FEEDS FOR AQUACULTURE: FREQUENTLY ASKED QUESTIONS*

National Oceanic and Atmospheric Administration

While all animals needs to eat and all farmed animals need to be fed, aquaculture represents the most efficient method by which to convert feed to edible protein. Research through the NOAA-USDA Alternative Feeds Initiative has accelerated progress toward reducing fishmeal and fish oil use in aquaculture feeds while maintaining the important human health benefits of seafood consumption. The remarkable progress in developing alternatives has reduced reliance on wild fish caught for this purpose.

This series of Frequently Asked Questions (FAQs) addresses commonly asked questions related to feeds used in marine aquaculture. We discuss what farmed fish eat and examine issues such as fishmeal and fish oil use in aquaculture and research efforts underway to bring greater sustainability to feed production.

1. What type of food
2. Do all fish
3. Why use fishmeal and fish
4. Where does fishmeal and fish
5. What are forage fish

---

* This is an edited, reformatted and augmented version of the following NOAA Fisheries website http://www.nmfs.noaa.gov/aquaculture/faqs/faq_feeds.html viewed July 2014.

6. What are fishmeal and fish
7. Does aquaculture
8. Doesn't harvesting
9. What is being done to develop alternative feed ingredients
10. What are potential alternatives to feeding fish
11. Can some of the waste
12. What about arguments that farmed fish
13. Are there growth
14. Are antibiotics fed to fish
15. Are antibiotics used at all in U.S. aquaculture
16. Are farmed salmon
17. How many pounds of other fish
18. Why don't we just eat fish
19. Doesn't uneaten fish
20. Does NOAA
21. What are some of the constraints to reducing aquaculture
22. What are specific areas of research that will lead

1. What Type of Food Do Farmed Fish Eat?

Farmed fish and shrimp eat feed that is specially formulated to contain all the essential nutrients they need to keep them healthy and growing and maintain the human health benefits of seafood consumption. The ingredients are formed into pellets, similar in many ways to dry dog food.

There are about 40 essential nutrients needed by all animals. Categories of essential nutrient include vitamins, dietary minerals, essential fatty acids and essential amino acids. These are provided by a number of feed ingredients including fish, plant, and processing waste meals and oils.

Did you know?

Farmed shellfish such as oysters, clams and mussels do not need to be fed a manufactured feed. These shellfish are "filter feeders" and consume plankton and other particles present in the water.

2. Do All Farmed Fish Eat the Same Thing?

No. Fish nutritional needs vary by species. Herbivorous fish eat a feed mixture that may contain plant proteins (e.g., soy, corn), vegetable oils, minerals, and vitamins. In the wild, carnivorous fish such as salmon eat other fish. Therefore, feeds for farmed carnivorous fish (as well as many herbivorous fish) include fish oils and proteins as well as plant proteins, minerals, and vitamins that achieve the nutrition requirements of the fish and offer health benefits to humans. Traditionally, diets for carnivorous fish contained 30-50% fish meal and oil; however, continued research is leading to greatly reduced reliance on these ingredients.

3. Why Use Fishmeal and Fish Oil in the Diets of Farmed Fish?

While fish and shrimp don't *need* fishmeal and oil in their diets, these ingredients have almost a perfect balance of the 40 or so essential nutrients that animals need to be healthy and grow – the same reason that seafood is so good for humans as well.

Fishmeal is a natural and well-balanced source of high-quality protein. As ingredients in aquaculture feed, fishmeal and fish oil supply essential amino acids and fatty acids reflected in the normal diet of fish. *Fish oil* is a major natural source of the healthy omega-3 fatty acids eicosapentaenoic acid (EPA) and docosahexaenoic acid (DHA). These fatty acids are not made by the fish, but become concentrated in fish further up the food chain from the marine phytoplankton (microscopic marine algae and microbes) that do synthesize them. Through research, we are learning that other combinations of ingredients can achieve the balance of the 40 essential nutrients. Affordable replacement ingredients for fishmeal and fish oil are becoming increasingly common, which is leading to declining percentage on those ingredients in farmed fish diets.

Including ingredients like oils from phytoplankton maintain the nutrient requirements of the final product without depending on fish oil. The economics of using blended oils is improving as fish oil prices rise and the technology to produce phytoplankton (and other replacement ingredients) improves.

Did you know?

Fish and shellfish—both wild and farmed—are very nutritious for humans. Fish are a natural and well-balanced source of high-quality protein. The unsaturated fats in marine fish, known as omega-3 fatty acids (eicosapentaenoic acid (EPA) and docosahexaenoic acid (DHA)) are proven to benefit heart and cardiovascular health. Omega-3-rich diets from oily fish such as sardines, herring, trout, and salmon are associated with a reduced risk of heart attack, stroke, and other cardiovascular diseases. Recent research has shown that Omega-3s may improve cognitive and neurological health as well. Marine fish contain significantly higher amounts of long-chain omega-3 fatty acids than terrestrial animals or freshwater fish. See our *Seafood & Health FAQ* for more information.

4. Where Does Fishmeal and Fish Oil Come from?

About ¾ of the fishmeal and oil are produced from the harvest of small, open-ocean (pelagic) fish such as anchovies, herring, menhaden, capelin, anchovy, pilchard, sardines, and mackerel. These fish have short life cycles and are capable of rapid reproduction and stock replenishment. The other ¼ is generated from the scraps produced when fish are processed for human consumption.

The United States is a small net producer of both fishmeal and fish oil. The largest U.S. fishmeal and oil production comes from menhaden caught in the east coast and Gulf, the second largest component of US production comes from fish processing trimmings produced in Alaska's seafood industry.

Did you know?

Fisheries used for the production of fishmeal and oil are often referred to as 'reduction' or 'industrial' fisheries because of the steps used to process the fish into meal and oil. Industrial fishing is a major component (20-30%) of global fisheries and a vital livelihood for many coastal economies. Reduction fisheries mainly are located off the coast of Peru and Chile, as well as in the North Atlantic, North Sea, and Baltic Sea. Chile and Peru account for about 40 percent of the global production. Careful fisheries management, including quota and catch limit systems maintains the sustainability of these fisheries over time.

Because these reduction fisheries are heavily regulated and the supply has remained relatively constant, increased demand for these fish increases the price for fishmeal and fish oil. As a result, alternatives are being sought for use in aquaculture feeds.

For more information on fishmeal and fish oil benefits, sustainability, production, and quality assurance, visit the website of International Fishmeal and Fish Oil Organization (www.info.net).

5. What Are Forage Fish Used For?

The small, open-ocean fish that are caught in industrial fisheries are consumed directly, used as bait in commercial and recreational fisheries and in lobster traps, used as ingredients in the feed for aquaculture, pigs, poultry, cattle, and pet food; and are converted into fish oil pills.

6. What Are Fishmeal and Fish Oil Used For?

Fishmeal and fish oil supply several major industries because they are natural ingredients of high nutritional value. While they have been major ingredients of swine and poultry feeds for many decades, a growing percentage of these resources have been used to manufacture aquatic feeds. This is due to the worldwide growth of aquaculture over the past two decades. Demand for fish oil in the supplement industry also is rising rapidly.

Fishmeal and oil are beneficial components of the diets of farmed fish. Because of this, aquaculture producers have been willing to pay more for these ingredients. As well, substitutes to fishmeal and oil have long been developed for terrestrial animal agriculture, while substitutes for aquaculture are just now coming online. With rising costs for fishmeal and oil, aquaculture producers are developing cost-effective, yet healthful alternatives.

Did you know?

Replacements and alternatives for fish meal in aquaculture feeds is an active and important area of research in which NOAA other federal agencies are heavily involved.

The *NOAA-USDA* is identifying alternative dietary ingredients that will reduce the amount of fishmeal and fish oil contained in aquaculture feeds while maintaining the important human health benefits of farmed seafood. Ultimately, the initiative will lead to the commercialization of alternatives for some species, which will result in reduced dependence on marine fish resources by feed manufacturers and seafood farmers worldwide.

For example, recent research has led to success in the use of the leftover *trimmings* from the fish processing industry. (Fish trimmings consist of the parts of the fish that are not converted into fillets or steaks during processing). Trimmings that used to be waste now can constitute a significant percent of the raw materials for fishmeal production.

7. Does Aquaculture Consume More Wild Fish Than Is Produced?

When aquaculture is considered as an aggregate industry, the answer is no. Globally, aquaculture uses about half a metric ton of wild whole fish to produce one metric ton of farmed seafood, meaning that aquaculture is a net *producer* of protein.

Feed conversion ratios (the amount of feed eaten by a fish related to the amount that fish provides for human consumption) vary among species, but farmed fish are far more efficient at converting feed than wild fish or other farmed animals such as cows and pigs.

For more information on fishmeal and fish oil benefits, sustainability, production, and quality assurance, visit the website of International Fishmeal and Fish Oil Organization (www.info.net).

8. Doesn't Harvesting Pelagic Fish Have a Detrimental Impact on the Food Chain and Other Animals That Depend on Them?

Forage fish serve a dual function of contributing to overall ecosystem biomass and supporting global food supply, the latter through both direct human consumption and providing feed for terrestrial farming and aquaculture. There is concern that ecosystem function is reduced at current catch levels; addressing this concern happens through fishery management rules (such as catch quotas) within individual capture fisheries.

Aquaculture is one of many end-uses for harvested forage fish, which also include providing fishmeal in diets of livestock, serving as baitfish for commercial and recreational fisheries, and – increasingly – as nutritional supplement and pet food ingredients. In the absence of aquaculture, these fish would be consumed by other industries.

The world supply of fish from pelagic fisheries has remained relatively constant over the past twenty years at around 6 million metric tons. These types of fish generally are capable of rapid reproduction and stock replenishment. Many pelagic fisheries are recognized as successfully regulated and many stocks are fished at levels below the biomass that achieves maximum sustainable yield. Careful fisheries management, including quota and catch limit systems, maintains the sustainability of these fisheries over time.

9. What Is being Done to Develop Alternative Feed Ingredients?

Future growth of marine finfish and shrimp aquaculture will need protein and oil sources greater than current fishmeal and fish oil production can satisfy. NOAA, in partnership with the U.S. Department of Agriculture (USDA), launched the NOAA-USDA Alternative Feeds Initiative in 2007 to accelerate the development of alternative feeds for aquaculture. The purpose of the Alternative Feeds Initiative is to identify alternative dietary ingredients that will reduce the amount of fishmeal and fish oil contained in aquaculture feeds while maintaining the important human health benefits of farmed seafood.

In addition to looking for substitutes, research is examining how farmed fish utilize feed, varying formulations, timing dietary needs with developmental stages, and other strategies to improve feed use efficiency.

Ultimately, the initiative will lead to the commercialization of alternatives for some species that will result in reduced dependence on marine fish resources by feed manufacturers and seafood farmers worldwide. The biggest challenges for researchers are to develop alternative ingredients that fish will eat, that supply the nutrition fish require to grow, and to make available alternative ingredients that are commercially viable. Current research, including that being done through the NOAA, is making great strides toward overcoming these challenges.

10. What Are Potential Alternatives to Feeding Fish to Fish?

Potential alternative include meals and oils from plants (the greatest source of protein and edible oil on earth), fish processing waste, yeast, bugs and other special meals, and even seaweed. Potential alternative ingredients already in use include soybeans, barley, rice, peas, canola, lupine, wheat gluten, corn gluten, other various plant proteins, yeast, insects and algae. Other sources that show great promise include waste from bio-energy and bio-plastic production and fish processing waste (trimmings). Farmed seaweed has significant growth potential as a source of food and fiber for both aquaculture feed and human consumption. Researchers have been successful in identifying alternatives that grow fish and help maintain the human health benefits of eating seafood.

Did you know?

The diet of farmed fish is currently not exclusively based on feeding fish to fish. We understand that herbivorous fish can eat a feed mixture that may contain plant proteins (e.g., soy, corn), vegetable oils, minerals, and vitamins. We also know that in the wild, piscivorous fish (such as salmon) eat other fish.

However you may be surprised to learn that even farmed piscivorous fish a great deal of the diet includes plant proteins, oils, minerals, and vitamins as long as they achieve the nutrition requirements of the fish. As research explores alternatives to replace fishmeal and oil in the diets of farmed fish, the percentage of fish meal and fish oil will continue to decrease while still providing the human health benefits of eating seafood. What do you call a vegetarian salmon?

11. Can Some of the Waste That Results from Processing Fish be Used as Fish Food?

Yes. The use of fish processing waste (trimmings) for the production of fishmeal and oil may be used. For example, discarded by-catch in 1994 was estimated to total 17.9-39.5 million metric tons (mt). This is a potentially large source of fishmeal. More recent estimates of by-catch of 20 million mt are equivalent to 25% of the reported annual harvest by marine capture fisheries. In Alaska alone, the fish processing industry generates about 1.2 million mt of by-product waste annually, from which high quality fishmeal and oil suitable for aquatic feeds may be produced.

Did you know?

Scientists at NOAA's Northwest Fisheries Science Center have been working on finding profitable uses for fish trimmings since the 1950s. Current research focuses on incorporating fish trimmings cost-effectively and within the constraints of short seasons, variable volumes of materials, and remote locations. NOAA scientists are working with industry academic partners to create a production process that can cost-effectively process the trimmings in to a high-value dried meal, suitable for feeding to numerous species in aquaculture. Preliminary fish growth studies have shown equivalent performance between meal using trimmings created with the new process and high quality conventional fish meal and fish oil. Other alternative proteins and oils from plants or other organisms can be used in combination with this high performance meal to achieve excellent fish growth.

12. What about Arguments That Farmed Fish Are Contaminated with Mercury and Other Heavy Metals?

No farmed fish are on the "avoid" list due to mercury. These compounds enter and concentrate in organisms largely through what they eat. Just like feeds for other domestic animals, aquaculture feeds are regulated by the U.S. Food & Drug Administration (FDA) and Departments of Agriculture in respective states, with advisement from the Association of American Feed Control Officials (AAFCO). The FDA and state agencies conduct inspections as well as collect and analyze feed and fish samples to help ensure that feeds and the fish that consume them meet strict state and federal requirements. Formulated feed ingredients used in aquaculture are regularly monitored to avoid possible contamination of feed with methyl mercury. According to the FDA and EPA, studies show that for people eating the standard U.S. diet, the health benefits of omega-3 fatty acids far outweigh the potential drawbacks of mercury toxicity due to fish consumption. Specific advisories are in effect for pregnant women, those trying to get pregnant, nursing women, and children.

13. Are There Growth Hormones in U.S. Farmed fish?

Growth hormone is not used in U.S. aquaculture. Although growth hormones may be given to other farm animals such as cattle and sheep, their use in food fish is prohibited by the U.S. Food and Drug Administration (FDA). Even if it was allowed, there is no incentive to use it, as it has not been shown to improve growth or efficiency in farmed fish. Certain additives such as pigments, antioxidants, and other nutritional supplements have been proven safe and their use in fish feeds is permitted by FDA regulation.

14. Are Antibiotics Fed to Fish to Improve Growth (or for Any Reason Other than Disease)?

In the United States, antibiotics are not fed to fish for non-therapeutic reasons through their feed or any other mechanism. The use of antibiotics for non-therapeutic purposes in aquaculture is prohibited by law. Incidentally, antibiotics do not improve growth or efficiency in fish (like they do in cows, swine, and chickens). However, antibiotics have been known to be added to fish food in other countries.

15. Are Antibiotics Used at All in U.S. Aquaculture?

As vaccines have been developed for the major diseases that impact aquaculture (including salmon), antibiotic use has all but disappeared in the

U.S. There occasionally is still a need to use them in special cases approved by a vet. All drugs, including antibiotics, used in aquatic species farmed in the U.S. have been proven safe and effective and are undetectable at the time of harvest (as prescribed by FDA withdrawal times). At present, only three antibiotics are registered and sold for use in the United States as feed additives for disease control in farmed fish. The use of parasiticides is similarly restricted by FDA regulations.

16. Are Farmed Salmon Fed or Injected with Dyes?

No. In the wild, salmon eat krill and other tiny shellfish that contain natural pigments called carotenoids, which are powerful antioxidants and precursors of vitamin A. Carotenoids, specifically astaxanthin and canthaxanthin, give salmon flesh its distinctive pigment. Farmed salmon are supplemented with natural and/or synthetic astaxanthin that is identical to the pigment that salmon get in the wild. Both natural and synthetic astaxanthin are processed and absorbed by wild and farmed fish in exactly the same manner.

17. How Many Pounds of Other Fish Are Required to Produce One Pound of Salmon?

None – there is no "requirement" for fish to be in diets for farmed salmon or any other fish. It is possible to feed salmon on a fish-free diet. However, inclusion of fish in the diets of farmed salmon reflects the diet of wild salmon and confers the nutrient benefits associated with eating salmon. Until replacements (such as algae) with nutritional elements similar to fishmeal and fish oil are commercialized, salmon and marine fish will need some fishmeal and fish oil in their diet. However, researchers are making significant progress in reducing the amount of fishmeal and fish oil used in the diets of farmed seafood while maintaining the human health benefits.

The answer to how much is used (as opposed to required) is a moving target, depending on how much fish meal and fish oil is used in any particular salmon diet. Recent data indicates reflects that the production of one pound of farmed salmon uses the fish oil from about five pounds of wild fish (depending on the oil content of the species used), but the fishmeal from only

1.3 pounds of fish. The excess fishmeal from those wild fish is used in the diets of other farmed animals.

> ***Did you know?***
> Aquaculture producers have been able to reduce the amount of fishmeal and fish oil in the diets of farmed fish by targeted specific growth periods. Increasingly, fishmeal and fish oil is used only for broodstock and larval nutrition, as well as in the finishing diet before the product goes to market. This minimizes the amount of fishmeal and fish oil used while maintaining the health of the fish during growth and the health benefits of the finished product.

Considered in the aggregate, aquaculture uses about half a metric ton of wild whole fish to produce one metric ton of farmed seafood. The amount of fish you get "out" (as seafood) relative to the amount of fish you put "in" (in the diet) – known as "fish-in/fish-out" (FIFO) conversion ratios – vary greatly among species. Piscivorous fish – especially those new to aquaculture – require more fishmeal and oil in their diet than omnivorous species.

18. Why Don't We Just Eat Fish from These Pelagic Fisheries (i.e. Further Down the Food Chain)?

To a certain extent, we do. Increasingly, species like mackerel, herring, sardines, and anchovies are sold for human consumption. As well, fish oil in the dietary supplement industry is the fastest growing segment of that market. However, consumer demand for species such as salmon, grouper, cod, and tuna drives the market for both farmed and wild fish. Pelagic fish (such as anchovy and menhaden) generally are in far less demand (especially in western countries) for direct human consumption.

> ***Did you know?***
> Site "fallowing" is a technique used to address potential environmental impacts. For example, state law in Maine requires salmon farms to "fallow" their sites for one year after harvest.

Fallowing requires the removal and cleaning of all equipment and leaving the site undisturbed, which allows the sea floor beneath the sea cages to recover from any potential impacts of fish farming.

19. Doesn't Uneaten Fish Food Accumulate on the Ocean Floor and Pose an Environmental Risk?

Jurisdictions that regulate net-pen aquaculture set limits to what can accumulate under aquaculture facilities. In the U.S., this is typically set at "no net accumulation" on an annual basis. Feed for farmed animals is the most expensive input to fish farming (about 60% of the cost of growing fish). For this and environmental considerations, fish farmers take extreme precautions to avoid over-feeding. Farmers use advanced technology including automated feeders and underwater cameras to provide feed and monitor consumption. Due to these advances in technology and careful siting, feed accumulation on the ocean floor is much less of a consideration than it once was. Additionally, fish farmers work with state regulators to monitor the benthic effects of their operations.

20. Does NOAA Fund Research on Alternative Feeds?

The National Oceanic and Atmospheric Administration (NOAA), in partnership with the U.S. Department of Agriculture (USDA), launched the NOAA-USDA Alternative Feeds Initiative in 2007 to accelerate the development of alternative feeds for aquaculture. The purpose of the Alternative Feeds Initiative is to identify alternative dietary ingredients that will reduce the amount of fishmeal and fish oil contained in aquaculture feeds while maintaining the important human health benefits of farmed seafood.

Since 1998, NOAA has been supporting alternative feeds research through the National Marine Aquaculture Initiative, a competitive grants program. Through this program, the agency has funded nutrition projects for a variety of marine fish species including black sea bass, cobia, cod, flounder, shrimp, snapper, sablefish and tuna. This research generated information on use of probiotics, identification of dietary requirements, and use of alternative proteins and processing byproducts. Since the 1950s, NOAA Fisheries Service labs have worked on diet development for Atlantic and Pacific salmon, sablefish, black sea bass, lingcod, rockfish, and several other marine species. NOAA labs helped develop methods for improved recovery and use of seafood processing waste and invasive species meals for use in aquaculture feeds.

21. What Are Some of the Constraints to Reducing Aquaculture's Dependence on Wild Fisheries?

Fish meal and fish oil produced from wild harvest fisheries are among the highest quality proteins and lipids available. For several decades, the supply has been stable but fully exploited; that is unlikely to change no matter if the demand is from aquaculture, agriculture, human consumption, or any other source. Although NOAA, its partners, and others are making significant strides toward reducing and, in some cases, replacing fishmeal and fish oil in aquaculture diets, two major constraints exist.

- *Growth in worldwide population and the related increase in demand for all types of food.* As landings from wild fisheries are limited to maintain healthy fish populations, the expanded supply of seafood will have to come from aquaculture fed on alternative feeds. While this is a constraint, it also is an opportunity for the United States to develop environmentally-responsible commercial aquaculture and work with partners to improve aquaculture practices around the world.
- *The technical challenge of alternative ingredients.* The challenges for researchers are to develop alternative ingredients that fish will eat and that supply the nutrition fish require to grow; and to make available alternative ingredients that are commercially viable. Current research, including that being done through the NOAA, is making great strides toward overcoming these challenges.

22. What Are Specific Areas of Research That Will Lead to Decreased Dependence on Fishmeal and Fish Oil?

The development of alternative feeds would benefit from increased investment in research and development in many areas. Key among these is to develop cost-effective ways to produce the hearth healthy marine omega-3 fatty acids. Continued evaluation of alternative protein and oil sources such as by-products from other industries (e.g., biofuel), fish, and newer sources such as worms, insects, and algae will yield insights into the nutrient profiles of alternative ingredients and how they may meet the nutrient requirements of farmed species.

INDEX

C

D

E

F

G

H

N

O

P

T

U

V